Illustrated BUYER'S ★ GUIDE™

FIREBIRD

Third Edition

John A. Gunnell

MBI Publishing Company

First published in 1998 by MBI Publishing Company, 729 Prospect Avenue, PO Box 1, Osceola, WI 54020-0001 USA

The information in this book is true and complete to the best of our knowledge. All recommendations are made without any guarantee on the part of the author or Publisher, who also disclaim any liability incurred in connection with the use of this data or specific details.

We recognize that some words, model names and designations, for example, mentioned herein are the property of the trademark holder. We use them for identification purposes only. This is not an official publication.

MBI Publishing Company books are also available at discounts in bulk quantity for industrial or sales-promotional use. For details write to Special Sales Manager at Motorbooks International Wholesalers & Distributors, 729 Prospect Avenue, Osceola, WI 54020-0001 USA.

Library of Congress Cataloging-in-Publication Data
  Gunnell, John
    Ilustrated buyer's guide. Firebird/John A. Gunnell. —3rd ed.
      p. cm.—(Illustrated buyer's guide series)
Rev. ed. of: Illustrated Firebird buyer's guide. 2nd ed. 1992.
ISBN 0-7603-0602-8 (alk. paper)
  1. Firebird automobile—Purchasing. I. Title. II. Series: illustrated buyer's guide series.
TL215.F57G85   1998
629.222'2—dc21   98-8040

On the front cover: A 1974 Firebird and a new 1998 Trans Am. *Tom Shaw*

On the back cover: A 1968 Firebird 350 HO coupe. *Pontiac Motor Division*

Designed by Bruce Leckie

Printed in the United States of America

Contents

Acknowledgments

Originally, I felt this guide could be finished in three months. It wound up taking almost three years. Other writers have called the story "fabulous." It is also fabulously complex. These cars are the ultimate expression of the auto-making philosophy put forth by General Motors' famous chairman, Alfred P. Sloan, Jr.

In 1959, Sloan said that through different "mixes" of colors, interiors, and optional equipment, General Motors could conceivably build a year's worth of cars with no two exactly alike. Models such as the Firebird took this concept to its highest stage.

Sophisticated enthusiasts can pinpoint thousands of subtle differences among Firebirds. The focus of this book is on those variations and their relative significance in the collector car field. What makes one car worth $300 and another $30,000? That is what this book will try to explain.

Initially, I felt that my previous research for *75 Years of Pontiac Oakland* (Crestline Publishing), *Classic Motorbooks Pontiac Trans Am Photofacts 1967–1973* (Motorbooks International Publishers & Wholesalers), and the *Standard Catalog of American Cars 1946–1975* (Krause Publications), combined with my experience with Old Cars Weekly and the Old Cars Price Guide (Krause Publications) would carry the project quickly to completion. But, I was wrong; in the end I had to rely on nearly 50 other people to unravel the mysteries of the Firebird marketplace.

These enthusiasts, with their personal knowledge, their cars, their photographic skills, and facts from books they have written were a big help in creating this guide. If you find it to be a useful tool in buying, selling, and collecting Firebirds, please join me in giving recognition and heartfelt thanks to the following experts: Bob Adams, Charles Adams, F. Ann Billiamosa, Stan Binnie, Tom Bonsall, Gerry Burger, Ted Cram, George Dammann, Larry DeLay, Dave Doern, Carl Dwiggens, Tom Gibson, Mike Green (Pace Publications), Cliff Gromer, Phil Hall, Barbara K. Harold, Reggie Harris, Jerry Heasley, Michael J. Hicks, Richard Johnson, Bill Kosfeld, Chet Krause, Mike Lamm, Larry Loernz, Jack Martin, Gary MacCoy, Skip McCully, Malcom McKellar, Ron McQueeney, Roy Nagel, Joe Oldham, Tim Parker, Bill Pearson, Mike Pergande, Dave Poltzer, Larry Pruchinak, Jerry Redden, John Sawruk, Carl Schaeffer, Marty Schorr, "Tex" Smith, Ken Steffensen, Dick Thompson, Tom Warth, Jay Wetzel, Gary Witzenburg, and Jill Witzenburg.

Also due a heap of thanks are the following companies: Automobile Quarterly Publications; Bookman Dan; Crestline Publishing; CSK Publishing (High-Performance Pontiac magazine); Indianapolis Motor Speedway Corporation (photographic department); Krause Publications; *Old Cars Weekly;* Pontiac Motor Division, General Motors Corporation; Quicksilver Publications; Redden's Relics; and, of course, MBI Publishing Company.

Introduction

Firebird began as Pontiac's answer to the Mustang. Like Mustangs, Firebirds have an enthusiast following. They seem to appeal to a smaller, more specialized group, but Firebird lovers are no less passionate than Mustang lovers in terms of their brand loyalty.

Despite their appeal, Firebirds have not gained a reputation as hot investment cars. Current prices seem very affordable, yet the marque offers above average potential for future value appreciation. While the days of the 200–300-percent increases that I mentioned in the first edition of this book are gone, marque experts predict Firebird prices will rise 25–50 percent in the next five years. That's on a par with or better than many other types of investments available today.

The rationale for such predictions is based on two factors. First, Pontiac experts are seeing many more Firebirds at shows these days than ever before. The 1998 St. Ignace Antique Car Show had dozens of the cars on display, including a 1967 coupe driven in the "Down Memory Lane" parade by its original owner—a 91-year-old woman—and a brand new Hurst Firebird driven by lovely Linda Vaughn, who has promoted Hurst performance products for years. Second, there are serious rumors about General Motors halting production of Camaros and Firebirds. Experts say that this action could send values of the two cars soaring.

As I predicted in the first edition of Illustrated Firebird Buyer's Guide, the cars went through the typical "12-year-value-increase" cycle around 1989. This is the point where cars begin to become collectible, partly because decay and deterioration reduce the existing supplies of a particular model. A study in the May 1978 issue of Car Collector magazine (now called Car Collector & Car Classics) illustrated that the rate of decay for the average car begins to lessen when it is 10 years old. Thus, this seems to be the point where collectibility starts to increase.

Classic car auctioneer Dean Kruse says that 20 years is the point at which a car's rate of value appreciation can be predicted to rise again. This is when many enthusiasts who liked certain cars when they were young have reached the point in their lives where they can afford to purchase restored examples of those same models. If this is true, some Firebirds that are now reaching their 20th anniversary—like the 1979 Daytona Pace Car—should see some excellent appreciation. So, as you read this edition, pay particular attention to these and other 1978–79 special-interest models.

This doesn't mean that all Firebirds that haven't "turned 20" are inexpensive or of little interest to collectors. The "Bandit" Trans Ams manufactured by Trans Am Specialties are an example of an aftermarket-converted model that generates extremely high collector interest and prices. The factory also produced a good number of special-interest Firebirds such as the 1983 Daytona 500 T-top 1984 15th Anniversary model, the 1989 Indy Pace Car, and the 1994 25th Anniversary coupe and convertible.

Prices on the early cars have risen since the first edition of this book was published. At that time, the early cars were just nudging the $10,000 range, SD-455s were pulling down just under $15,000, and offers in the $20,000 range might have bought one of eight 1969 Trans Am convertibles that Pontiac built. According to the August 1998 edition of Old Cars Price Guide, the base price for a 1967 Firebird coupe in top shape is now a firm to $10,500, and ragtops start at $1,000 more. The guide says to add 15 percent for a 350 HO engine,

10 percent for a four-speed transmission, and 30 percent for the Ram Air 400 model option. The 1973 Trans Am SD-455 coupe with four-speed transmission is up to $18,400, and the 1969 Trans Am ragtops are up to $26,400– $28,600, depending on equipment options.

Such prices represent the current maximums for Firebirds (and if you ask me, I think they're a bit conservative). However, they do not reflect "average" price trends. Many "ordinary" vintage Firebirds can still be purchased at prices between $1,500 and $5,000 (for vehicles in good-to-excellent conditions). In addition, many bargains still show up among the rarer, early models, because the idiosyncrasies of the Firebird market are so little understood by non-specialized buyers and sellers.

Growth of the Firebird parts supply business and increased club activity are additional signs that the collectibility of Firebirds is on the upswing. The economics of manufacturing upholstery kits, body repair panels, trim parts, and mechanical items demand that a sizable market for these components exist before suppliers start to tool up for production. Within the past three or four years, reproduction Firebird parts have begun to appear regularly in the catalogs of many suppliers. Meanwhile, the number of Firebird owners joining such organizations as the Pontiac-Oakland Club International and the Trans-Am Club of America has noticeably increased.

The number of books published about a particular marque is another indication of its collectibility. As recently as 1979, Classic Motorbooks' catalog listed only three Firebird references: a magazine article and two tune-up manuals. Only the magazine article was geared mainly to collectors. In contrast, Classic Motorbooks' latest advertisements list over a dozen books devoted to Firebirds. Nine of them concentrate on the collector market.

While general interest in Firebirds is growing rapidly, it's important to note that there are different types of Firebirds, plus a variety of equipment and trim packages. These factors have an immense influence on the values of the cars and the desirability of different models. Since its introduction, the Firebird has been marketed as a car line that came in several models and body styles, each including specific features at a given base price. Options and accessories could be ordered at extra cost, allowing the buyer to build the exact type of car he or she desired.

For the first three years of production, 1967 to 1969, two body styles were offered—coupe and convertible. Either could be purchased in any of five models, resulting in 10 separate configurations available. And the "building block" concept expanded this option total even further.

With the late introduction of 1970 Firebirds (often referred to as 1970 1/2 models), the convertible was deleted. The number of individual models was reduced to four. However, there was still a wide choice of extra-cost equipment and trim. This allowed each customer to tailor the car to personal tastes and driving habits.

Today, Firebirds with certain engines and transmissions, interior appointments, paint colors, trim packages, and so on can be worth two to five times as much as other Firebirds of the same vintage. Values are also significantly affected by the type of body style, the choice of different models, and the relative scarcity of the overall combination of features and equipment. It takes a lot of detailed knowledge to be a wise Firebird buyer.

Let's talk about the Firebird as an investment. Can you expect to buy one, fix it or drive it, and someday realize more from its sale than you have invested in it?

If you pay the right price for your Firebird today, the answer is an unqualified yes! You'll recoup your original purchase price, your investment in restoration and maintenance, plus hidden costs like storage fees and inflation.

This statement is based upon consideration of the two factors—popularity lag and life cycle—which indicate that pre-1982 Firebird values will probably zoom in the near future.

This book covers over 185 different types of Firebirds, which is a great deal of ground to cover. The only way to accomplish this is to dispense with long, drawn-out discussions of Firebird history. (For those interested in Firebird history, I suggest *The Fabulous Firebird* by Michael Lamm and *Firebird! America's Premier Performance Car* by Gary L. Witzenburg.) Instead, this book concentrates on year-to-year changes, model identification, factory identification codes, strong and weak points of different models, production data, driving qualities, comparative performance, and how to appraise a car's fair market value. This is the information you'll want at your disposal when considering the purchase of a Firebird.

Investment Rating

Collector car investments are a function of supply versus demand. Rare models with the most buyers seeking them generally bring the highest levels of price appreciation in the open market. Such cars usually represent the best investments.

Using original production totals as a guideline, the relative supply of a given car model can be estimated. Some researchers have considered available statistics on automobile scrap rates in given years to pinpoint how many examples of a specific model should still be around. Use of one method or the other depends upon how "scientific" you want your estimate to be, but both make it possible to consider the supply factor at least somewhat objectively.

The demand for a given model is a more subjective consideration. Different observers have various opinions on which Firebird models are in high, average, or low demand. (For example, the buyer who wants a common model just like the one he or she used to own, will approach a purchase in a high demand state of mind.) Opinions about demand given in this book are based on my research and contacts with hundreds of Firebird enthusiasts.

Each model or specific option package covered here was entered on a master sheet. The star rating system was then employed to rate each model's investment potential relative to that of other Firebirds.

I have avoided using half-star ratings because of my personal experience in using another type of five-class rating. It's my belief that modifications to a five-class rating reflect subtle differences between models that are recognized by the experts, but have little direct affect on prices, values, or investment potential.

Rating systems are designed to enhance a researcher's objectivity and take the compiled data one step away from emotion. They also simplify the understanding of the data. On the other hand, using half stars reduces objectivity, takes the compiled data one step away from emotion, and creates confusion.

Limiting ratings to five stars did mean giving base Firebirds, Firebird Sprints, and Firebird 350s the same ratings in some cases. This doesn't mean that the three models are worth the same, but indicates that they will appreciate in value at approximately the same rate.

It's important to understand that the stars

Star code	Rating	Supply	Demand
★★★★★	Excellent	Very limited	Highest
★★★★	Very good	Below average	High
★★★	Good	Average	Above average
★★	Fair	Above average	Average
★	Poor	High	Low

reflect only investment potential. They do not pertain to values or prices. For instance, a newer Firebird might sell for more than an older one. On the other hand, it might not give you the same return on investment as the older Firebird.

For example, a newer Firebird bought for $14,000 today could be worth $9,000 five years from now. But an older Firebird purchased for $1,000 today, could be worth $5,000 in five years.

In this case, the older car represents the better investment, although the newer one is more valuable.

I hope this guide will help you select a Firebird that's worth purchasing and putting some money into. But, please do not use these ratings exclusively. The market can change and the predications can turn out to be wrong. The stars are a guide to direct your investments, but always get additional opinions before spending your own hard-earned cash.

Chapter 1

1967

The Firebird was introduced by Pontiac Motor Division (PMD) on February 23, 1967. It came as a coupe or convertible. Five basic engines were available: two overhead cam (OHC) six cylinders, two 326-ci V-8s, and one 400-ci V-8. Each powerplant was the starting point for a separately merchandised model. There was a Ram Air induction option for the Firebird 400. The Firebird 400 Ram Air was not considered a separate model.

Pontiac promoted the Firebirds as the Magnificent Five. Sales literature asked, "Which Firebird is for you?" This was directly opposed to the marketing approach Ford took with the Mustang, which stressed the idea of one basic car with a wide range of options. On each Firebird model, specific regular production options (RPOs) were considered standard equipment.

At their core, all Firebirds were Mustang-type cars: four-passenger sports/personal compacts. Pontiac referred to them as sports cars, stressing their differences, rather than their basic similarities. This reflected PMD General Manager John Z. DeLorean's original desire to market a true, two-seat sports car. General Motors' (GM) management nixed this concept and spun the production Firebirds off the Chevrolet Camaro.

Camaros and Firebirds used the same body shell. The Camaro made its debut in the fall of 1966. It took PMD five extra months to tailor its version of the car. Different sheet metal was used for the hood, grille, and taillight panels. Firebirds also had Pontiac powerplants and a chassis tuned by PMD engineers. Introduction of the line was made in midyear at the 1967 Chicago Automobile Show.

The base Firebird was at the low end of the scale. It was offered as a coupe ($2,666) and a convertible ($2,903). Bucket seats, vinyl upholstery, nylon-blend carpets, woodgrain dash trim, and E70-14 wide oval tires were standard. The engine was a regular-fuel OHC six with monojet carburetor. It developed 165 brake-horsepower at 4,700 rpm.

All 1967 Firebirds were available with standard and Custom interior trim options. The Custom option included slim-line bucket seats, molded interior door panels, integral front armrests, deluxe steering wheel, dashboard passenger-assist grip,

Although the Pontiac Firebird was based on the Chevrolet Camaro, it had a firmer suspension and distinctive front end. The sports compact was merchandised in five model-options: Firebird, 326, 326 HO, Sprint, and 400. Judging from engine call-outs on hood bulge, this is a Sprint convertible. *Pontiac Motor Division*

and decorative exterior trim moldings. A thin-profile bench seat was available as an option in any Firebird coupe.

An inflatable, Space-Saver spare tire was standard in all 1967 Firebirds. A conventional spare tire was a no-cost option.

For exterior identification, the base Firebird six had "3.8 liter: overhead cam" lettering on the rear edge of the bulge on the hood. The Custom trim package option added bright moldings on the rocker panels and wheel lip moldings.

The next model up the scale was the Firebird Sprint, featuring all the basics plus heavy-duty suspension, a floor-mounted gear shift, and a hotter OHC six. This engine featured a racier cam and a four-barrel carburetor, and it produced 215 horsepower at 5,200 rpm. The price of the Sprint equipment was $116.16 extra.

On the exterior, the Sprint models had the same lettering on the rear edge of the hood bulge. There were also OHC-6 Sprint badges on the forward edge of the rocker panels.

V-8 models started with the Firebird 326, which was less expensive than the Sprint six. This small-displacement V-8 equipment package retailed for $95 over base price. It included a two-barrel, 326-ci V-8 with 250 horsepower at 4,600 rpm that ran on regular gas. As with base Firebirds, a three-speed transmission with column-mounted shifter was standard.

The 326 V-8 also used lettering on the rear edge of the hood bulge for model identification. In this case, the engine call-outs read "326". Like the six-cylinder models, the cars with small V-8s had standard E70-14 tires.

The fourth Firebird was called the 326 HO. This was a high-output (HO) version of the standard 326, with a four-barrel carburetor and higher-compression (10.5:1) cylinder heads. The 326 HO also featured performance-type dual exhausts. The powerplant developed 285 horsepower at 5,000 rpm. Other standard equipment included the column-shifted three-speed manual transmission, heavy-duty suspension, heavy-duty battery, and F70-14 wide oval tires. The price for a 326 HO Firebird was $180.58 extra.

A distinctive feature of the 326 HO was its racing stripes on the side of the body. At the front, these stripes were separated by HO decals to identify the model.

Last among the models was the Firebird 400 with a 325-horsepower four-barrel V-8 linked to a heavy-duty three-speed manual transmission with a floor-mounted shifter. A distinctive twin-scoop hood characterized this high-performance model. Its price of $274 extra also included dual exhausts, heavy-duty battery, Power-Flex fan, heavy-duty suspension, and redline or whitewall tires.

The 1967 Firebird had one-year-only front ventipanes, on both coupes and convertibles. This is a 326 convertible owned by Tom Beck of Pennsylvania. The original red- or white-stripe tires have been replaced with white-letter style. *Old Cars Weekly*

Another view of the Firebird Sprint convertible shows 3.8-liter OHC hood call-out lettering more clearly, as well as Custom trim option upholstery design. The unique six-cylinder engine came in base (165 horsepower) and Sprint (230 horsepower) versions. *Pontiac Motor Division*

The $263 Ram Air package, available for the 400 model, had chrome engine parts and functional hood scoops. The scoops directed cold air to the four-barrel carburetor. According to Pontiac specs, this did not increase horsepower; instead, it raised the car's rpm peak to 5,200.

On its exterior, the Firebird 400 featured the special dual-scoop hood. The 400 engine call-outs were found on the sides of the scoops and the right-hand side of the deck lid. The 400 Ram Air had functional scoops.

All Firebirds used a conventional chassis layout with the engine in front, transmission in the middle, and drive axle at the rear.

Body construction was a uni-body structure connected to a ladder-type subframe that carried the front sheet metal, powertrain, and running gear. The two sections were bolted together at the front seat and cowl areas.

Coil-spring suspension was used up front, with mono-leaf springs at the rear. All models had a 108-inch wheelbase and a 188.4-inch overall length. Front and rear trends were 60 inches.

Although generally upgraded from Camaro specs, the base Firebird's rear suspension (with its mono-leaf springing) was prone to wheel hop on

Close-up of Sprint six convertible reveals the OHC six call-out badge on rocker panel. (Some preproduction publicity photos show a Sprint ragtop without this trim; don't be fooled!) Car shown here also has options including hood tach, radio (note antenna mounting), and Custom wheel covers. Tape stripe is another extra and is not the same as the one used on the HO model-option. *Pontiac Motor Division*

Here is the 1967 HO convertible, a restored example, at the Elkhart Lake Trans Am Territory. Notice the distinct bodyside tape stripe with HO letters incorporated in forward section. The rear-mount antenna and wire wheel covers are extras. The tires, however, again vary from stock design.

hard acceleration and braking. While the spring setup gave a fairly quiet ride, the overall ride quality of loaded cars was on the poor side.

Mid-range models had one adjustable track bar added to the right-hand side of the axle. Its job was to help fight wheel hop. High performance models also had a left-hand track bar, for even better traction, which gave them a very hard and firm ride.

The 1967 Firebird exhibited a predictable degree of understeer when driven hard, however, directional control was fair to good. The car's handling was generally quite quick and responsive, but the brakes were a problem. Manual brakes seemed to require extra effort to apply, while the optional power disc brake setup required too little effort. Fade characteristics were fair.

Firebirds offered a seating arrangement that was typical of contemporary pony cars: They were adequately roomy in front and tight-fitting at the rear; driving position was good, but not conducive to long-haul use. In a survey, owners rated the comfort level fair to poor for passengers in the rear.

Magazine editors road tested several 1967 Firebirds. Among them were two 400 convertibles with optional four-speeds and 3.90:1 axles. They did 0–60 miles per hour in 5.5 and 5.8 seconds, respectively, with quarter-mile runs of 14.4 seconds at 100 miles per hour, and 14 seconds at 104 miles per hour. *Car and Driver* noted a top speed of 114 miles per hour.

Motor Trend tested a 400 convertible with Turbo Hydra-matic and a 3.08 axle. Performance

Here is an example of the Firebird 400 convertible, which was characterized by twin-scoop hood with 400 engine call-outs. The 400 model-option was available in base 400 and Ram Air 400 editions; the main difference was the engine. Both put out 325 horsepower, but the "hotter" Ram Air I developed its power peak at lower rpm. (Car in background is 1954 turbine concept car, built by GM, which is also known as "Firebird.") This ragtop illustrates the standard interior trim. *Pontiac Motor Division*

was charted at 7.5 seconds for 0–60, and 15.4 seconds at 92 miles per hour for the quarter mile.

Road and Track put a Sprint six through its paces. Equipped with an optional four-speed transmission and standard 3.55:1 axle, the car went 0–60 miles per hour in 10.1 seconds and did the quarter-mile in 17.2 seconds at 81 miles per hour.

Consumer Reports described Firebird engine and performance characteristics as "smooth and quiet." Highway acceleration reserve for the six was described as "adequate" and for the V-8, "ample." The standard two-speed automatic (not used in 400s) was called "smooth-shifting, but not versatile."

The OHC six was said to have a few characteristic weaknesses and to be difficult to service because of its complexity. Most frequently cited as problems were a tendency to throw timing belts and a propensity for the distributor to skip time. PMD vigorously denied such problems, but *Consumer Reports* rated the engine "undesirable" in its 1971 buyer's guide issue. One documented problem was insufficient throttle linkage clearance, which could cause sticking in the wide-open position.

Other problems associated with 1967 Firebirds showed up in owner surveys. They included above-average repair rates for the fuel system and manual transmission on V-8 models.

Most collectors are able to identify different 1967 Firebirds through model features and engine characteristics or details. The cars also carry vehicle identification number (VIN) plates, which are attached to the left-front door pillar. These plates carry a code with 13 characters. The first character, 2, identifies the car as a Pontiac. The second and third characters, 23, indicate the model, Firebird. The fourth and fifth characters reveal the body style code (37 for coupes; 67 for convertibles). The sixth symbol identifies the model year, and the seventh designates the assembly plant (all 1967 Firebirds were made at the same plant in Lordstown, Ohio). The last six symbols are the sequential production number.

Five "models" were promoted by PMD's marketing arm, but the engineering department officially recognized only two kinds of Firebirds, the coupe (model 22337) and the convertible (model 22367). One of these two codes was used as the first five symbols in the VIN plate code for all 1967 Firebirds, followed by the year code (7 for 1967) and factory code (L for Lordstown). These seven symbols were then followed by the sequential numbers, which fell into two different ranges—one for sixes and one for V-8s.

Engine	Body style	Code
Six	Coupe	223377L600001 and up
Six	Convertible	223677L600001 and up
V-8	Coupe	223377L100001 to 600000
V-8	Convertible	223677L100001 to 600000

PMD's engineering specifications section also listed four different model codes on its Cars Built

Race car driver and builder, John Fitch, designed an aftermarket GT kit for the Firebird. A typical Fitch Firebird included suspension modifications, fast-ratio power steering, improved brakes, four-speed gearbox, air scoops, exhaust headers, aluminum deep-sump oil pan, low back-pressure exhaust system, oversize radial tires, long-range headlights, headlight flashers, and other features for about $1,000 over base Firebird price. *Car Exchange*

The Firebird 400 coupe, and all other 1967 models, came in 15 colors. This one is Cameo Ivory (color code D). Rally II mag-style wheels were a very popular accessory on all Firebirds of all years. *Old Cars Weekly*

Data Sheet. This is the only document I have ever seen showing breakouts by standard and deluxe designations. (These designations do not correspond to the five models marketed. They are probably related to the type of factory interior —standard or custom—that a car left the assembly line with). The engineering specifications section kept its production records according to these designations, rather than by "model" or body style. The totals were as follows:

Model	Name	VIN	Manual	Auto.	Total
67-23	Std.	223	5,258	5,597	10,855
67-24	Std.V-8	223	8,224	15,301	23,525
67-25	Del. Six	223	2,963	3,846	6,809
67-26	Del. V-8	223	11,526	29,845	41,371
	Total		27,971	54,589	82,560

PMD also kept 1967 Firebird production totals according to body style. These figures were as follows:

Body style	Total
Two-door hardtop (coupe)	67,032
Two-door convertible	15,528
Total	82,560

Engine identification codes were stamped on the cylinder block. Firebird OHC six codes were located on the right side of the block at the rear of the distributor. Firebird V-8 codes were on the front, right side of the block, just below the cylinder head. The engine block codes were as follows:

How do the 1967 Firebirds rate as investment cars? They will probably provide an average or slightly less than average return on investment over

Displacement	Engine	Carburetor	Horsepower	Transmission	Code	Compression
230 cid	ohc six	1-bbl	115	manual	ZK	9.0:1
				automatic	ZN	
	ohc Sprint six	4-bbl	215	manual	ZD	10.5:1
				automatic	ZE	
326 cid	326 V-8	2-bbl	250	manual	WP	9.2:1
				automatic	YO	
	326 HO V-8	4-bbl	285	manual	WR	10.5:1
				automatic	YP	
400 cid	400 V-8		325	manual	WU	10.75:1
				automatic	YE	
	400 Ram Air		325	manual	WZ	
				automatic	YF	

A first in the automobile industry was this new hood-mounted tachometer available on 1967 Pontiacs. Mounted in front of the windshield in line with driver viewing, this instrument permitted easy instant reading with a minimum of distraction.

Spyder section of Rally IIs was finished in dark argent (gray) to offset bright metal fasteners. Tires with thin, red or white stripes were standard equipment. *Pontiac Motor Division*

the next few years. That boils down to appreciation of 10 percent, or less, on an annual basis.

The 1967 models do have a couple of things going for them. First, they were the original Firebirds. Second, the supply of these cars is low because of the number of years since they were new. Third, they offer a few truly distinct features. They are the only Firebirds built with vent windows, 230-ci OHC sixes, and 326-ci V-8s.

On the other hand, they are the least technically sophisticated models in the Firebird car line.

Their suspensions are the weakest, and their drivetrains have a few characteristic bugs. Their styling is among the most radical for the Firebird series. People tend to either love them or hate them.

If your heart is set on owning a 1967 model, shop around for the best example you can find of one of the following variations: 326 HO convertible, 400 convertible, or 400 Ram Air coupe or convertible. Knowledgeable enthusiasts pick these three models as the 1967 Firebirds that offer a better-than-average investment opportunity.

1967 Firebird Options and Accessories

Sales Code	Description	Retail Price ($)
382	Door edge guards	na
631	Front floormats	na
632	Rear floormats	na
502	Power brakes w/pedal trim	41.60
501	Power steering (17.5:1 ratio)	94.79
551	Power windows (only 2,283 cars)	100.05
401	Luggage lamp	na
402	Ignition switch lamp	na
421	Underhood lamp	na
582	Custom air conditioning	355.98
374	Rear window defogger/blower	21.06
531	All Soft-Ray glass	30.54
532	Soft-Ray windshield	21.06
441	Cruise control (automatic only)	52.66
442	Safeguard speedometer	na
654	Fold-down rear seat (carpetback)	36.86
544	Power convertible top	52.66
474	Electric clock	15.80
394	Remote-control OSRV mirror	na
391	Visor vanity mirror	na
572	Headrest (bench or bucket seat)	52.66
342	Push-button radio w/manual antenna	61.09
341	Manual rear antenna (in lieu of front)	9.48
344	Push-button AM/FM radio w/manual antenna	133.76
351	Rear speaker	15.80
354	Delco stereo tape player	128.49
—	Stereo multiplex adapter	na

Sales Code	Description	Retail Price ($)
462	Deluxe steering wheel	na
471	Custom sports steering wheel	na
504	Tilt steering wheel (n.a. w/std. steering, 3-speed column shift or Turbo Hydra-matic w/o console)	na
731	Safe-T-Track differential	na
491	Rally stripes	na
444	Rally gauges (& hood tach 702)	31.60
704	Hood tachometer	63.19
524	Custom shift knob	na
472	Console (w/bucket seats & floor shift)	47.39
481	Dual exhaust (standard w/HO, 400)	30.23
481	Tail pipe extensions	na
621	Ride & Handling pkg. (standard w/HO and 400)	9.32
431	Front and rear custom seat belts	na
434	Front seat shoulder belts	na
521	Front disc brakes (502 recommended)	63.19
—	Hood retainer pins	na
494	Dual horns (standard w/Custom trim)	na
738	Rear axle options	n/c
OBC-SVT	Vinyl roof	84.26
OBC	Custom trim option	108.48
OBC	Strato-Bench front seat	31.60

Sales Code	Description	Retail Price ($)
684	H-D fan (base, Sprint)	8.43
514	H-D thermo fan (326, HO)	15.80
674	H-D alternator (standard w/A/C)	15.80
681	H-D radiator (standard w/A/C)	14.74
361	H-D dual-stage air cleaner	9.43
678	H-D Battery (included w/674)	3.48
453	Rally II wheels (available w/521)	
	Custom trim	55.81
	Std. trim	72.67
454	Rally I wheels (available w/521)	
	Custom trim	40.02
	Std. Trim	56.87
461	Deluxe wheel discs	na
458	Custom wheel discs	na
452	Wire wheel discs	69.51
—	Tu-tone	
	Standard colors	31.07
	Custom colors	114.27
—	185R-14 radial tires (whitewall)	
	400	10.53
	others	42.13
—	E70-14 redline or whitewall tires	
	400	n/c
	others	31.60

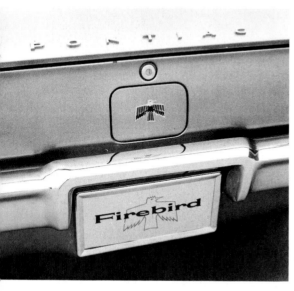

First Firebird had a much sturdier rear bumper than the 1967 Camaro, and a conventional fuel filler door, which Camaro omitted. It carried stylized Firebird badge, adopted from American Indian symbolism. *Pontiac Motor Division*

The Sprint option added a four-barrel carburetor on the OHC six. By the way, enthusiasts use the words "coupe" and "hard top" interchangeably when talking about non-convertible Firebirds. *Pontiac Motor Division*

Chapter 2

1968

Coupe	Convertible	
★	★★	Base Firebird
★	★★	Firebird Sprint
★	★★	Firebird 350
★★	★★★	Firebird 350 HO
★★★	★★★★	Firebird 400
★★★	★★★★	Firebird 400 HO
★★★	★★★★	Firebird Ram Air 400

One-piece side-glass, suspension improvements, and more powerful engines were features of Firebirds built in the series' second year. Production jumped to 90,152 coupes and 16,960 convertibles.

PMD sales literature said engineers "labored long and lovingly on a host of refinements which make the Magnificent Five even more so for 1978." In addition to the five models, HO and Ram Air options were available for the Firebird 400. A hotter Ram Air II engine was introduced in March 1968.

With the elimination of vent windows, flow-through ventilation was adopted. Given the trade name Astro-Ventilation, the new system added chrome-plated louver balls (similar to air-conditioning vents) on each lower corner of the dash. Other interior changes included redesigned door panels, use of expanded vinyl seat inserts (with Custom Trim), and new burled woodgrain trim on the centers of the dash and console.

To combat wheel hop, all models above base level had multi-leaf springs and staggered shocks at the rear. Wider F70-14 tires were standard, except on base Firebirds.

The 1968 Firebird drivetrains were upgraded with larger sixes, larger small-displacement V-8s, new V-8 cylinder heads, and several new high-performance options. There were electrical improvements, more standard safety equipment, and new options.

External styling alterations were minimal, but they were noticeable because of the lack of vent windows. The front parking lights wrapped around the body corners to double as safety side-markers. Rear side-markers, in the form of the PMD logo, were also added.

Prices for base Firebirds were $2,781 for coupes and $2,996 for convertibles. This included all GM safety equipment, front bucket seats, vinyl upholstery, woodgrain trim, outside rearview mirror, E70-14 blackwall tires, Space-Saver spare, and the 175-horsepower OHC six.

This new six had the same 3.88-inch bore, but its stroke was increased from 3.25 to 3.53 inches to give a 250-cubic inch (4.1 liter) displacement. A new monojet carburetor was used with stick-shift cars and helped reduce exhaust emissions. While previous OHC sixes had used the same crankshaft and connecting rods as a Chevrolet six, the 1968 engine had all Pontiac parts. A new four-speed transmission option was added.

Again priced at $116.16, the Sprint package added a floor-mounted shifter, OHC six Sprint emblems, body sill moldings, F70-14 tires, a four-barrel carburetor, and 10.5:1 heads. Horsepower for the 250-ci Sprint six remained the same (215) as for the 230-ci version, at the same rpm peak. A different, optional four-speed transmission was offered for 1968 Sprints and the three-speed automatic was dropped.

Retailing for $106 extra, another package including a small V-8 was named the Firebird 350. The new engine, based on the previous year's 326, featured a bore-size increase from 3.72 to 3.88 inches, while the stroke remained at 3.75 inches. It was rated for 265 horsepower at 4,600 rpm. Although the small-block heads again gave 9.2:1 compression, they were redesigned with deeper, smoother combustion chambers and larger, more upright valves to promote a controlled combustion environment. They were part of the PMD's emissions program, as was a new, thermostatically controlled carburetor preheating system.

The 350 package also had such features as a two-barrel carb; F70-14 tires; and a column-shifted, manual three-speed transmission. A heavy-duty three speed, four-speed, and two-speed automatics were additional transmission choices at extra cost.

For $181 over base price, a 350 HO package was the starting point for high-performance models. In addition to (or instead of) the above features, it included a four-barrel carburetor, dual exhausts, HO side stripes, heavy-duty battery, and F70-14 tires.

This engine gave 320 horsepower at 5,100 rpm. The rather large jump (35 horsepower) was not due entirely to more cubic inches. A high-output camshaft, better exhaust-gas scavenging, and a different carburetion setup made this a potent powerplant for its size.

The basic 400 package was $351 to $435 over base (lower with Turbo Hydra-matic or four-speed transmissions). Among its ingredients were a floor-mounted shifter; chrome engine parts; sports suspension and shocks; heavy-duty battery; dual exhausts; special emblems; dual-scoop hood; red-stripe or whitewall tires; and Power-Flex fan.

Three 400-ci engines (4.12x3.75-inch bore and stroke)—all with 10.75:1 heads—were available at first. The W66 version produced 330 horsepower at 4,800 rpm. The L74 HO version had a wilder cam and gave 335 horsepower at 5,300 rpm. Then came the L67 version, featuring cold-air induction through functional hood scoops. It was also rated 335 horsepower, but at a lower (5,000) rpm peak.

Cars with 400 HO and Ram Air 400 engines were tested by two magazines. The L74 went 0–60 in 5.3 seconds and did the quarter-mile in 14.2 seconds (*Car and Driver,* March 1968). With the L67, these times changed to 4.8 seconds and 15 seconds, respectively (*Sports Car Graphic,* February

New for 1968 was ventless side window treatment. Shown here is the 350 HO convertible—a restored version wearing white-letter tires, but otherwise totally stock. *Old Cars Weekly*

This base version of the Sprint convertible showed up at the Carlisle Flea Market Car Corral. Stripes on these whitewalls are slightly wider than originals, but a good modern replacement. *Old Cars Weekly*

1968). Top speed for both was 110 miles per hour at the end of its quarter mile run.

Both of these cars used what Pontiac buffs call '67 heads. These had 2.11-inch intake valves and 1.77-inch exhaust valves. The Ram Air II (also coded L67)—an engine made available around April 1968—came with a new type of round-port head.

Changing the head configuration gave 36 percent more breathing room for greatly improved airflow characteristics. In addition, the $631.12 option included a high-lift cam; forged aluminum pistons; Armasteel crankshaft; new pushrods and guides; tulip-head valves; and dual, high-rate valve springs. It was rated 340 horsepower at 5,400 rpm. The option did not include a Turbo Hydra-matic or four-speed manual transmission, either of which was mandatory for Ram Air II cars, at extra cost.

Many Firebird engineering features seemed to be improved for 1968, but the cars still received low ratings from *Consumer Reports*. The magazine again described both sixes and V-8s as "undesirable cars."

Sixes again recorded above-average incidences of mechanical engine repairs. In addition, owner surveys showed a higher level of repair work was put into bodies, electrical systems, exhaust systems, and shock absorbers. Some of this could have been due to heavy-duty use, such as in street racing.

Visual and trim differences between models were much the same as the previous year. The base six had hood lettering changed to read "overhead cam: 4.1 liter," instead of "3.8 liter: overhead cam." Sprints again had rocker panel identification badges. Base V-8s had 350 call-outs on the hood. Side stripes with HO lettering characterized 350 HOs. All of the big-block cars had 400 lettering on their hood scoops. Engine features and decals could aid identification efforts, too.

In 1968, vehicle identification plates were moved to the left, upper instrument panel surface, as part of an industrywide antitheft effort. They were visible from outside the cars through the windshields. Also starting in 1968, VINs were stamped on all engines and transmissions. This is an aid to collectors who want to ensure that a car has its original powertrain.

The basic coding system was similar to (and equally as confusing as) that used the previous year. A typical 1968 Firebird code looked like this: (model) 8 (L) 600001. Model codes for both coupe and convertible body styles were

Nicely equipped Firebird 350 coupe wears options including hood tachometer, wire wheel covers, vinyl top, and rear-mounted radio antenna. Note front parking lights that wrap around side of body to double as safety side-marker lamps. *Pontiac Motor Division*

unchanged. The sixth digit, denoting model year, was an 8 for 1968. Factory codes and starting sequential numbers were similar to those in 1967.

PMD Cars Built Data Sheets, issued by the engineering specifications section, again showed four model suffixes for standard and deluxe six and V-8 models. The VINs given on the sheets corresponded to those suffixes. Using those designations, production counts were as follows:

Model	Name	VIN	Manual	Auto.	Total
68-23	Std. Six	223	7,528	8,441	15,969
68-24	Std. V-8	224	16,632	39,250	55,882
68-25	Del. Six	225	1,216	1,309	2,525
68-26	Del. V-8	226	7,534	25,502	32,736
	Total		32,910	74,202	107,112

Despite the various engineering refinements made in 1968 Firebirds, there are no strong indications that they should be rated as generally better investments than 1967 models. In fact, 1968 Firebirds seem to have spent more time in repair shops than the first-year cars. This indicates that they would be slightly more troublesome to owners who plan to drive them frequently.

Models of the two years look virtually identical. Their distinctive styling won't appeal to everyone and makes a preference in this area purely a matter of personal choice. Those who like their undulating lines, peaked and bulging hoods, and "snouty" appearance will probably prefer these cars over later models, which are generally considered to have cleaner designs. On the other hand, those who don't like the early Firebird look are probably in the majority and unlikely to change their preference. In most cases, a more radically styled car will have a smaller following as far as total numbers go. This seems to be true with the Firebirds of 1967 and 1968.

The hot models of both years—350 HO, 400 HO, and Ram Air 400—are true high-performance machines, some of which are very rare. For example, only about 2,500 400 Ram Airs were

Rear side-marker lights, in the form of a red V-shaped Pontiac emblem, were new for 1968, under government safety regulations. Shown here is the Firebird 350 HO coupe. There was also a 400 HO for 1968, which carried 400 lettering on the passenger side of the rear decklid. *Pontiac Motor Division*

Displacement	Engine	Carburetor	Horsepower	Transmission	Code	Compression
250 cid	ohc six	1-bbl	175	manual	ZK	9.01:1
				automatic	ZN	
	ohc Sprint six	4-bbl	215	manual	ZD	10.5:1
				automatic	ZE	
350 cid	350 V-8	2-bbl	265	manual	WC	9.2:1
				automatic	WJ	
	350 HO V-8	4-bbl	320	manual	WK	10.5:1
				automatic	YM	
400 cid	400 V-8		330		YW	10.75:1
					YT	
	400 HO V-8		335	manual	WI	
				automatic	XN	
	400 Ram Air V-8		335		WQ	
					WZ	

produced 1967, and the 1968 total for this model was certainly not much higher. Cars with the Ram Air II option also seem to be especially rare. However, none of these cars have yet attracted the degree of appeal that collectors lavish on Trans Am convertibles or 1973–74 Trans Ams and Formulas with the Super Duty (SD) 455 engine. It's impossible to give the early performance models the top, five-star rating, when a good example can be purchased for half the price of, say, a 1969 Trans Am ragtop—and will also appreciate at a slower rate.

The same is true when the five-star rating system is used to compare the 1968 base Firebird coupe with a Sprint coupe, or a Firebird 350 coupe with a Sprint coupe or Firebird 350 coupe of the same vintage. For the one model year, the higher-level model would, logically, be the better investment. But when some 167 different 1967–85 models are compared, a degree of generalization must be used. This trends to de-emphasize the subtle differences among models in a given year.

In the open market, when cars are sold at auctions or through classified ads in enthusiast publications, relatively few, highly specialized models stand out as the cars in which to invest. Among the 1968 Firebirds, the standouts would be the Ram Air 400 coupe, and the 400, 400 HO, and 400 Ram Air convertibles. In addition, either a coupe or a convertible with the Ram Air II package

The Custom trim interior option for 1968 had wide, vertical pleats. These were made of expanded vinyl. *Pontiac Motor Division*

Rear passenger compartment offered tight passenger accommodations with three sets of seat belts. A folding rear seat was optional and cost $42.13 extra. *Pontiac Motor Division*

represents the ultimate model of this year worth sinking money into. These cars will appreciate much more rapidly than all of the other 1968s.

If you are interested in making a solid investment on the low side of the price scale, then your choice should be any of the six-cylinder or small V-8 convertibles. With such a car, your investment won't double or triple overnight, but appreciation will still be much higher than the average of 10 percent annually.

If you are looking for a "sleeper"—a model that doesn't seem especially desirable but may become more so because it's unusual or rare—the model 68-25 Deluxe Six may be the answer. The problem is that it's difficult to figure out exactly what this designation means. PMD's engineering specifications section seems to be the only branch of the company that used the standard and deluxe nomenclature (to differentiate the interior trims).

However, the production figures show that only 2,525 of these cars were made, of which only 1,216 had manual transmission. That would make survivors quite rare and, possibly, just weird enough to be great long-term investments. After all, buying low and selling high is part of the auto investment game.

I assume that a Firebird Deluxe Six is a base or Sprint model with the Custom interior. If this is correct and you could find one built as a convert-

ible with equipment like the Sprint package and a four-speed, chances are that you would have yourself a nearly one of a kind Firebird.

Standard Firebird interior featured seats with thinner, multiple horizontal ribs on either side of center seam. Door panels also varied in design from Custom trim. Also available as an option was Pontiac's Astro bench-type seat with folding center armrest. This was the heyday of bucket seats with consoles, and few buyers chose the bench seat.

Twin hood scoops gave the 400 Ram Air Firebird immediate eye appeal. This is the 400 Ram Air convertible with Rally II wheels and rear antenna. *Pontiac Motor Division*

1968 Firebird Options and Accessories

Sales Code	Description	Retail Price ($)
591	Speedometer gear adapter	11.59
731	H-D air cleaner	9.48
582	Custom air conditioner	360.20
701	H-D battery (standard w/4-bbl; six w/o 582)	15.80
474	Elec. clock (not available w/394)	15.80
514	H-D clutch & fan, 7-blade (w/V-8, w/o 582)	15.80
	(w/V-8, w/o 582)	15.80
472	Console (n.a. w/contour bench seat)	50.55
441	Cruise control (n.a. w/manual trans.)	52.66
492	Remote-control deck lid	13.69
404	Rear window defogger (except Conv.)	21.06
361	Safe-T-Track differential	42.13
361	H-D Safe-T-Track (w/H, K, P, S axles)	63.19
521	Front disc brakes	63.19
342	4-bbl ohc Sprint six (w/o 582)	116.16
343	2-bbl 350 V-8	105.60
344	4-bbl 350 HO V-8	180.58
345	4-bbl 400 V-8	
	w/351, 354, 358	273.83
	w/o 351, 354, 358	358.09
347	4-bbl 400 Ram Air V-8	
	(w/351, 354, 358; n.a. w/582)	616.12
348	4-bbl 400 HO V-8 (w/351, 354, 358)	350.72
481	Dual exhausts (w/343)	30.54
482	Exhaust extensions	
	w/341, 342, 343	10.53
	w/481, 344, 345, 347, 348	21.06
444	Auxiliary gauge cluster	31.60
531	All tinted glass	30.54
532	Tinted windshield	21.06
412	Door edge guards (2-door)	6.24
571	Contoured head restraints	52.66
414	Dual horns	4.21
524	Custom gear shift knob (man./floor)	4.21
671	Underhood lamp	3.16
672	Ignition switch lamp	2.11
652	Luggage lamp	3.16
631	Front mats	6.85
632	Rear mats	8.43
732	Trunk mat	8.43
422	RH and LH visor vanity mirror	4.21
421	RH visor vanity mirror (exc. w/422)	2.11
424	LH remote-control OSRV mirror	9.48
SPS	Special solid paint (exc. Code A, black)	83.20
STT	Special two-tone paint (coupe)	114.80
RTT	Two-tone paint (std. color on coupe)	31.60
SPR	Special paint (code A, black only)	10.53
502	Power brakes	42.13
561	Full-width 4-way power bench seat	69.51

Sales Code	Description	Retail Price ($)
501	Power steering	94.79
564	LH 4-way power bucket seat	69.51
544	Power convertible top	52.66
551	Power windows	100.05
381	Manually operated rear antenna	9.48
394	Stereo tape player (n.a. w/391, 392)	133.76
382	Push-button radio and antenna	61.09
384	Push-button AM/FM radio & man. antenna	133.76
391	Rear seat speaker	15.80
494	Rally side stripes (n.a. w/344)	14.74
431	Front & rear Custom seat belts	9.48
432	Rear seat shoulder belts (w/431, 754)	26.33
754	Front seat shoulder belts	
	w/431	26.33
	w/o 431	23.17
568	Contour bench seat	31.60
604	Rear folding seat	42.13
634	Adjustable front & rear shock absorbers	52.66
442	Safeguard speedometer	10.53
621	Ride & Handling package (w/o 345)	9.48
621	Ride & Handling package (w/345)	4.21
471	Custom sport steering wheel	
	w/554	30.54
	w/o 554	45.29
462	Deluxe steering wheel	14.74
504	Tilt wheel (n.a. w/3-speed column shift or Hydra-matic w/o console; power steering required)	42.13
434	Hood-mounted tachometer	63.19
402	Spare tire cover	5.27
SVT	Cordova vinyl top (coupe)	84.26
351	Turbo Hydra-matic (w/345, 347, 348)	236.97
352	Automatic w/341, 342, 343, 344)	194.84
354	4-speed w/floor shift (w/o 37S axle)	184.31
355	H-D 3-speed w/floor shift	
	w/343, 344	84.26
	w/345, 348	n/c
356	H-D 3-speed w/floor shift (use w/472; incl. w/342)	42.13
358	Close-ratio 4-speed w/floor shift (mand. w/37S)	184.31
534	Custom pedal trim (na w/554)	5.27
458	Custom wheel discs	
	w/554	20.01
	w/o 554	52.66
461	Deluxe wheel discs	21.06
452	Wire wheel discs	
	w/554	52.66
	w/o 554	73.72

Sales Code	Description	Retail Price ($)
453	Rally II rims	
	w554	63.19
	w/o 554	84.26
321	Front foam cushion	86.37
554	Custom trim option*	114.88
332	Lamp group	5.25
331	Mirror group (exc. conv.)	13.22
322	Rear window defogger-protection group	55.25
THC	E70-14 red-stripe tires (w/341 engine)	31.60
THD	E70-14 white-stripe tires (w/341)	31.60
THE	E-70-14 blackwall tire (w/341)	n/c
TKM	195R-14 whitewall rayon radial tires (n.a. w/461)	
	w/341 engine	56.87
	w/342, 343, 344	42.13
	w/345, 347, 348	10.53

Sales Code	Description	Retail Price ($)
TMC	F70-14 redline nylon tires	
	w/341 engine	46.34
	w/342, 343, 344	31.60
	w/345, 347, 348	n/c
TME	F70-14 blackwall nylon tires	
	w/341 engine	14.74
	w/342, 343, 344	n/c

*Custom trim option includes dual horns, deluxe wheel discs, deluxe steering wheel, custom trim pedal plates, front & rear wheel opening moldings, roof rail moldings, windshield pillar garnish moldings and Custom seat, door and quarter panel trim, and assist bar.

Chapter 3

1969

Coupe	Convertible	
★	★★	Base Firebird
★	★★	Firebird Sprint
★	★★	Firebird 350
★★	★★★	Firebird 350 HO
★★	★★★	Firebird 400
★★★	★★★★	Firebird 400 HO
★★★★	★★★★	Firebird Ram Air IV 400
★★★★	★★★★★	Trans Am Ram Air III
★★★★	★★★★★	Trans Am Ram Air IV

There were basically five Firebirds—base, Sprint, 350, 350 HO, and 400—available again for 1969. The 400 model could be ordered with HO or Ram Air options. An exciting, new Trans Am model was introduced in mid-season. Like the other Firebirds, it came in coupe or convertible styles.

All 1969 models had obvious styling revisions. The new design marked a departure from the past and gave a glimpse toward the future. However, it was more of a major facelift than a radical change. There were very few, very minor, technical updates.

Retailoring of the outer sheet metal was based on changes in the Camaro. New characteristics included a slightly lower roofline, flatter wheel openings, sculptured bodyside panels, and front fenders with windsplits. The front end featured a new, integral bumper grille made of space-age Lectern plastic. It had a Mercedes-inspired split grille treatment with dual headlamps set into square, individual openings.

The 108-inch wheelbase and 60-inch track were continued, but overall length increased to 191.6 inches. All models were about one inch

The big news for 1969 was the midyear introduction of the Trans Am. Only eight Trans Am ragtops were ever assembled. This one belongs to Charles Adams, an executive with Pontiac Motor Division's advertising agency. *Ted Cram, Contemporary Historical Vehicle Association*

wider and 60 pounds heavier.

Thirty-three safety, antitheft, or convenience features were standard in all 1969 Firebirds. They included front shoulder belts (in coupes), front-seat headrests, uniform automatic-transmission quadrants, and improved fuel tank retention provisions.

Interiors received lots of attention, starting with wider, softer front bucket seats and camera-case-grain instrument panels. There were five standard trims and nine custom options, including a rare, gold leather combination. Bench seats were optional for coupes, but rarely ordered. A steering column ignition lock was new.

Optional Custom (vinyl) trim packages included breathable knit-style upholstery, bright roof-rail interior moldings, woodgrain dash trim, a molded trunk mat, integral front armrests, padded door panels, a passenger assist grip, and assorted interior and exterior trim items.

Base models were the coupe ($2,831) and the convertible ($3,045), which again could be identified by hood call-outs reading "overhead cam: 4.1 liter." The Sprint six package was $121 extra. Red Sprint badges appeared on the rocker panel moldings. Both cars had the expanded assortment (as in GM basics), plus the same equipment distinctions of previous years.

Red 350 call-outs on the sides of the hood bulge identified 1969 Firebirds with the small V-8.

This package retailed for $111 over base prices. Some minor gear box ratio changes were about the only revisions from 1968 specifications.

Side stripes were no longer used on, or available for, the Firebird 350 HO. This gave a cleaner look, but made it impossible to distinguish cars with the $186 package externally. Decals on the 350 engine were the main clue to HO models.

The 350 HO was one powerplant that PMD modified for 1969. Changes included new cylinder heads; a higher-lift, longer-duration cam; larger-diameter valves; and smaller carburetor bores. As a result, horsepower climbed to 325 at 3,600 rpm.

Scoops on either side of the hood bulge, decorated with red 400 emblems, made it easy to pick out the Firebirds in the big-block series. The package carried prices between $275 and $358, depending on the type of transmission.

Except for different weights and the deletion of an optional 2.56 ratio axle, specs for the regular 400 were unchanged from 1968. The same was true for the 400 HO package, which added a high-output cam and freer-flowing exhaust system for about $77 more. (The hotter cam was used only with cars having manual transmission attachments. The cam used with Turbo Hydra-matics had the same specs as the base 400 edition.) The 400 HO engine was also called the Ram Air III engine.

Trans Am convertible production included four cars with four-speeds and four with Turbo Hydra-matic. This rare car, belonging to Charles Adams, is in good original condition, but tires don't match.

They are white-stripes on front, white-letters on back. This, of course, wasn't the way they came. *Ted Cram, Contemporary Historical Vehicle Association*

Ram Air IV was PMD's ultimate powertrain option for Firebird 400s. It added some $832 to base model prices. The ingredients included a high-output cam and springs, oversize valves, and flat-top pistons with valve indents. Rated at 345 horsepower at 5,400 rpm, the Ram Air IV was good for 14-second quarter-mile runs with terminal speeds of 101 miles per hour.

Despite continuing improvements and refinements, *Consumer Reports* still rated Firebirds with six-cylinders and V-8s as "undesirable" models from a used-car buyer's point of view. The magazine said that the six had slightly above-average repair rates; it required more repairs than a 1967, but less than a comparable 1968. The V-8 models required the average number of repairs—an improvement over the earlier Firebird V-8s.

This was the lowest production year for OHC six Firebirds; they accounted for only 14 percent of total production. Owner surveys showed that the 1969s were more mechanically reliable than previous OHC sixes and required less body and shock absorber repairs. However, electrical and exhaust systems still gave owners some problems.

The same owner surveys showed that 1969 Firebird V-8s required more attention to bodies and electrical systems performed with more reliability than in the past.

Specific body and mechanical drawbacks that knowledgeable Firebird buyers look for are rust in the lower rear corners of the front fenders, rotten rocker panels and doglegs, loose timing chains, broken shocks, and corroded wiring connections. Many 1969 Firebirds offered for sale today—especially ragtops—have been quickly repaired with fiberglass body filler, which will not provide permanent protection against rust and corrosion.

The first Trans Am made its debut on March 8, 1969, at the Chicago Automobile Show. It was a Firebird loaded with specific options selling for about $1,100 (prices varied per transmission and body style) over base price. The WS4 Trans Am Convenience and Performance package was available only this one year on either the coupe or convertible. Convertibles were built in extremely limited numbers.

Originally intended as a sedan-racer that would be eligible to compete in the Sports Car Club of America (SCCA) Trans-Am road racing series, the production model became something else. This was because a 303-ci, regular-fuel racing engine never got to the assembly line. Instead, the model used one of two different 400-ci V-8s, which had displacements too large for SCCA racing. PMD made a royalty arrangement whereby the cars could be called Trans Ams, even though they could not race in the series. This agreement cost the company $5 per car but helped generate interest in the new model—although many sports car enthusiasts did not favor the misuse of the name.

The Trans Am option included all base Firebird equipment, plus a special hood, blacked-out grille, three-spoke woodgrained steering wheel, Quadrajet carburetor, three-speed heavy-duty manual transmission with floor shift, chrome engine parts, free-flowing dual exhaust system, rear deck airfoil, heavy-duty springs and shocks, front and rear stabilizers, Power-Flex fan, Safe-T-Track differential, full-length body stripes, front disc brakes, variable-ratio power steering, F70-14 whitewall glass-belted tires, and functional front-fender air extractors.

The base Firebird came with the base version of the OHC six. The 1969 taillights were wider, Pontiac lettering was moved down to rear end panel, and gas door disappeared. Smaller Firebird symbol, painted red, incorporated the trunk lid lock.

Base engine for the Trans Am was the 400 HO (Ram Air III) with 335 horsepower. The Ram Air IV powerplant was optional. Only one color scheme—cameo white with blue stripes—was offered for the Trans Am. Interiors came in black, blue, or parchment. Identification features included Trans Am and Ram Air decals.

An especially rare and desirable model is the 1969 Trans Am convertible, of which only eight examples were built. All of them used the 400 HO engine; four had manual transmissions, four had Turbo Hydra-matics. Convertible tops were available in blue or white. One of the eight cars was recently advertised in the Orb Report (a prestige collector magazine) with a high, but not unrealistic, asking price of $35,000. Production of 1969

The Sprint six coupe, last of the first-generation Firebirds, was heavily restyled along contemporary Pontiac lines with integral bumper-grille, integral headlamps, squarish sheet metal, and a touch of sculpturing. *Pontiac Motor Division*

Top photo from 1969 catalog shows the standard Firebird interior features. Center panel shows optional deluxe steering wheel with three vinyl-padded spokes. Bottom details illustrate the hood tachometer, front fender windsplits, and custom wheel covers. *Tom Gibson*

Firebirds moved very sluggishly. Only 88,405 units were turned out for the model year, which was extended to 16 months to help clear out dealer overstocks. When other 1970 Pontiac models were introduced in the fall of 1969, there was no new Trans Am. Instead, introduction of an all-new type of Firebird was put off until mid-season, and units built to 1969 specifications continued to be sold through the end of the calendar year. (Some of these cars may have been titled as 1970 models.)

Serial numbers for 1969 Firebirds were again located on the left-hand side of the instrument panel. The coding system was the same as that used in 1968. Model numbers were again identified by the first five VIN characters: 22337 for coupes and 22367 for convertibles. The sixth symbol was a 9 for 1969 (carryover cars built after January 1, 1970 had a 0). The seventh symbol was an L for the Lordstown, Ohio, factory. The last six numbers were production sequence numbers, ranging from 100001 to 600000 for V-8s and from 600001 and up for OHC sixes.

Production records were kept for all Firebirds as a group, and for Trans Am coupes and convertibles separately. In addition, there were breakouts for Firebirds according to transmission type, and for Trans Ams according to engines and transmissions. It is not known how many Firebirds were built in each of the five basic models. The available records look like this:

Name	Coupe	Convertible	Total
Firebird	75,362	11,649	87,011
Trans Am	689	8	697
Total	76,051	11,657	87,708

Name	Manual	Automatic	Total
Firebird and Trans Am	20,840	66,868	87,708

Trans Am Engine	Body style	Manual	Automatic	Total
Ram Air IV (L67)	coupe	46	9	55
Ram Air HO (L74)	coupe	114	520	634
	convertible	4	4	8
Total		164	533	697

The 1969 Firebird continued to carry two-letter engine codes stamped in the same places on six- and eight-cylinder blocks. The codes were as follows:

Displ.	Engine	Carburetor	HP	Transmission	Code	Compression
250 cid	ohc six	1-bbl	175	manual	ZC	9.0:1
				automatic	ZF	
		4-bbl	230	manual	ZH	10.5:1
			235	automatic	ZL	
350 cid	350 V-8	2-bbl	265	manual	WM	9.2:1
				automatic	XB	
					YE	
	350 HO V-8	4-bbl	325	manual	WN	10.5:1
				automatic	XC	
400 cid	400 V-8		330	manual	WZ	10.75:1
				automatic	YT	
	400 HO V-8 (Ram Air III)		335	manual	WQ	
				automatic	YW	
	400 Ram Air IV		345	manual	WH	
				automatic	XN	

Do 1969 Firebirds represent a good buy for today's enthusiast or collector? The answer is definitely yes! Among Firebirds of the first three years, these cars seem to be the most popular. And, despite their modest production, they appear to be available in relatively good supply.

The hot number to own is the Trans Am; the first of the breed and a true, limited production car. It goes without saying that the convertible represents one of the ultimate Firebird investments. However, chances of finding an authentic Trans Am convertible for sale are nearly nonexistent. Beware of forgeries, as several seem to be around.

Trans Am coupes, Ram Air IV-equipped Firebird 400s, and 400 HO ragtops are all very good investments. Supplies of such models are well below average, but demand is very high. The Trans Am coupe with the Ram Air IV engine and Turbo Hydra-matic is another true rarity, although most collectors would probably prefer a four speed.

Other 1969 models compare to their 1967–68 counterparts on the broad scale, but keep in mind that use of the five-star rating system comes with built-in limits on flexibility. On a one-to-one basis, the 1969 versions of a given model would seem to be a better investment compared to earlier editions. And these cars are also seen and traded much more frequently, which means that the market is there.

Another consideration is the minor, but nevertheless real, technical refinements made for 1969. These may not add up to an extra star on a chart rating relative factors for over 160 models, but they are something to think about when toying with the idea of purchasing a specific car.

Also note that that the production of manual-transmission cars declined a bit in 1969. A stick-

Trans Am name appeared in small, blue lettering on the front fender ahead of wheelwell. A pilot model shown in one advertisement had similar red lettering on the spoiler, but it did not appear on regular production cars. *Dave Doern*

Front fender air extractor scoops were on both front-fenders, behind wheel openings. They were functional. *Dave Doern*

Trans Am package was most clearly characterized by twin air-slot hood, longitudinal body stripes, front fender air extractors, and rear decklid airfoil or spoiler. Grille had flat blacked-out finish. *Dave Doern*

shift car has slightly more intrinsic value to the enthusiast. With rarity tossed in for good measure, a four-speed manual gearbox could have a somewhat positive effect on future values. That could make cars with this feature slightly better investments. In 1969, Pontiac used two different types of codes to identify different options and accessories. Both had three symbols in most cases. The first was a sales code that dealership personnel used for ordering equipment. The second was the GM Universal Production Code (UPC) used by the assembly line workers who installed the equipment. Most collectors are more familiar with the UPC, but both are provided here:

1969 Firebird Options and Accessories

Sales Code and UPC	Description	Retail Price ($)
582/C60	Custom A/C	
	(n.a. w/Sprint or Ram Air)	375.99
731/K45	H-D dual-stage air cleaner	9.48
682/K82	H-D 55-amp. alternator	
	(w/341, 342)	15.80
688/K96	55-amp. alternator	
	(V-8 only; n.a. w/582)	
	w/343, 344	26.33
	w/345, 347, 348	15.80
672/UA1	H-D battery	
	(std. w/344, 345, 347, 348, 582)	4.21
474/U35	Elec. clock (incl. w/484; n.a. w/444)	15.80
472/D55	Console (w/bucket seats)	53.71
SVT/G08	Cordova top (coupe)	89.52

Sales Code and UPC	Description	Retail Price ($)
441/K30	Cruise control (with auto., V-8 only)	57.93
492/A90	Remote-control decklid release	14.74
404/C50	Rear window defogger	
	(n.a. w/convertible)	22.12
481/N10	Dual exhausts (avail. w/343)	30.54
691/K02	H-D fan w/fan clutch	
	(n.a. w/341, 347, 348, 582)	15.80
692/KB2	H-D Power-Flex fan	
	all w/o 582	10.53
	V-8 w/582	n/c
534/M09	Gear knob (floor shift/man. trans.)	5.27
532/A02	Tinted windshield	22.12
531/A01	All tinted glass	32.65
412/B93	Door edge guards	6.24

The Trans Am had regular Firebird interiors. Car shown here has the Custom trim option, custom dash and custom door panels, and also the sport steering wheel. Only 127 Trans Ams had manual gearboxes like this one. The Trans Am option could not be teamed with the bench seat available in other Firebirds. *Gerry Burger*

Sales Code and UPC	Description	Retail Price ($)
414/U05	Dual horns	4.21
694/V64	Instant-Aire pump (n.a. w/347, 611)	15.80
652/U25	Luggage lamp	3.16
671/U26	Underhood lamp	4.21
631/B32	Front floor mats	6.85
632/B33	Rear floor mats	6.32
424/D33	OSRV remote-control mirror	10.53
421/D34	RH visor vanity mirror	2.11
422/DH5	LH visor vanity mirror	2.11
SPO/-	Special solid colors	
	coupe w/o vinyl top	115.85
	conv. or coupe w/vinyl top	100.05
RTT/-	Standard two-tone color (except convertible)	31.60
STT/-	Special two-tone color (exc. conv.)	147.45
SPR/-	Special color codes F, T, W	12.64
514/JL1	Custom pedal trim plates	5.27
501/N41	Variable-ratio power steering	105.32

Sales Code and UPC	Description	Retail Price ($)
564/A46	LH power bucket seat	73.72
544/C06	Power convertible top	52.66
551/A31	Power windows	105.32
502/J50	Power brakes (n.a. w/345, 347, 348)	42.13
511/JL2	Power front disc brakes	64.25
588/C57	Power flow ventilation (n.a. w/582)	42.13
701/V01	H-D radiator (n.a. w/582)	14.74
382/U63	AM radio w/manual antenna	61.09
384/U69	AM/FM radio w/manual antenna	133.76
388/U58	AM/FM stereo radio w/man. antenna	239.08
484/W63	Rally gauge cluster w/clock (w/o 442, 444)	47.39
444/U30	Rally gauge cluster & tach. (w/o 442, 474, 484)	84.26
621/Y96	Ride & Handling springs & shocks	
	w/345, 347, 348	4.21
	w/341, 342, 343, 344	9.48
604/A67	Rear folding seat	42.13

Sales Code and UPC	Description	Retail Price ($)
431/WS1	Custom front seat belts, includes shoulder straps	
	Coupe	12.64
	Convertible	36.86
432/WS2	Custom seat belts, includes front & rear shoulder straps	
	Coupe	38.97
	Convertible	63.19
438/AS1	Front shoulder straps (conv. only)	10.53
708/N65	Space-Saver spare tire	
	w/o 454	15.80
	w/454	n/c
402/P17	Spare tire cover (n.a. w/708)	5.27
504/N33	Tilt steering (n.a. w/std. steering or column-shift manual transmission)	45.29
391/U80	Rear speaker (n.a. w/388, 394)	15.80
442/U15	Safeguard speedo. (n.a. w/484, 444)	11.59
461/N30	Deluxe steering wheel	15.80
462/N34	Custom sports steering wheel	
	w/324	34.76
	w/o 324	50.55

Sales Code and UPC	Description	Retail Price ($)
394/U57	Stereo tape player (n.a. w/391)	133.76
471/UB5	Hood-mounted tachometer	63.19
554/W54	Custom trim option	
	w/knit-vinyl bench, coupe	110.99
	w/bucket seats	78.99
	w/leather bucket seats	199.05
452/P02	Custom wheel covers	
	w/324	20.01
	w/o 324	41.07
451/P01	Deluxe wheel cover	21.06
453/N95	Wire wheel covers	
	w/324	52.66
	w/o 324	73.72
454/N98	Rally II wheels	
	w/324	63.19
	w/o 324	84.26
321/Y88	Basic group	113.75
324/Y86	Décor group	62.14
331/WS6	Power-assist group	
	w/341, 342	364.93
	w/343	375.49
	w/344, 345, 347, 348	396.61

Base engine for the Trans Am sometimes is referred to as the 400 HO V-8, but enthusiasts prefer the name Ram Air III. It came standard with chrome dress-up parts. There was also a Ram Air IV extra-performance option. *Dave Doern*

Sales Code and UPC	Description	Retail Price ($)
332/WS5	Turnpike cruise group (w/HM, V-8, 501)	176.94
334/WS8	Rally group	
	all 345, 347, 348; w/324	149.55
	all 345, 347, 348; w/o 324	186.41
	all 341, 342, 343, 344; w/324	154.82
	all 341, 342, 343, 344; w/o 324	191.68
342/W53	Sprint sport option (w/o 582)	129.54
343/L30	Firebird 350 sport opt.	110.59
344/L76	Firebird HO sport opt.	199.05
344/W66	Firebird 400 sport opt.	
	Coupe w/351, 354, 358	347.56
	Convertible w/351, 354, 358	331.76
	Coupe w/o 351, 354, 358	431.81
	Convertible w/o 351, 354, 358	416.01
322/WS4	Trans Am option (w/354 only)	724.60
341/STD	ohc six 1-bbl	n/c

Sales Code and UPC	Description	Retail Price ($)
348/L74	400 V-8 4-bbl Ram Air (w/345)	
347/L67	400 V-8 4 bbl Ram Air IV (w/345)	558.20
361/G80	Safe-T-Track axle	
	regular	42.13
	H-D	63.19
364/G95-G97	Economy axle	2.11
368/G90-G92	Performance axle	2.11
362/G94-G83	Special order axle	2.11
359/M38	Turbo Hydra-matic	
	w/341, 342	195.36
	w/343	205.92
351/M40	Turbo Hydra-matic (w/344, 345, 347, 348)	227.04
352/M31	Two-speed automatic	
	w/341	163.68
	w/343	174.24

Apparently, there was a factory variation on application of stripes at the rear. Most cars had them under the spoiler like this. Some cars had them running over the spoiler. Rear decklid latch panel was finished in blue. *Dave Doern*

Sales Code and UPC	Description	Retail Price ($)
353/Std.	Three-speed manual column (w/341, 343, 344)	n/c
354/M20	Four-speed manual, floor shift (w/all exc. 347, V-8; w/3.90, 4.33 axles)	195.36
355/M13	H-D 3-speed manual, floor shift (w/343, 344)	84.26
356/M12	Three-speed manual, floor shift	
	w/341	42.13
	w/342	n/c

Sales Code and UPC	Description	Retail Price ($)
358/M21	Close-ratio 4-speed, floor shift (w/390)	195.36
HR/PL3	E78-14 white tire (w/341)	26.33
GF/PX5	F78-14 black tire	14.74
GR/PX6	F78-14 whitewall tire	28.44
MT/PY5	F70-14 redwall tire	74.78
MT/PY4	F70-14 whitewall tire	74.78

Chapter 4

1970

In 1970, Pontiac's Firebird went through a complete metamorphosis. The 1967–69 models had been designed to compete for sales against Mustang-class cars. But these "second-generation" Firebirds were aimed at imported sports cars and, in the Trans Am's case, the Corvette.

PMD promoted the new Firebirds as the "beginning of tomorrow." It was a perceptive slogan, since the basic design lasted 12 years. During that span, there were outstanding high-performance models, many exciting special trim packages, and some truly collectible limited-edition Firebirds. But, for 1970, the biggest changes were in styling and the restructuring of car lines. The four new models were just the start of many better things to come.

Pontiac stylists must have worked overtime on the fabulous new Firebird body that appeared at the Chicago Automobile Show on February 26, 1970. It was extremely clean and smooth-lined, with a semi-fastback roof, a body-color Endura rubber nose, single headlamps, and a smartly tapered tail.

Available only as a hardtop coupe, these 1970 1/2 Firebirds, as they are generally referred to, came in four base packages: Firebird, Esprit, Formula 400, and Trans Am. All of them retained the 108-inch wheelbase and overall length of 191.6 inches. Front track was increased to 61.3 inches, while rear track remained 60 inches.

While styling was the strong point of the 1970 1/2 models, some evolutionary engineering changes were made as well. They included the addition of standard front disc drakes, redesigned front suspension, improved steering linkage, better

body sealing, double-panel roof construction, additional acoustical insulation, and new multi-leaf rear springs.

Pontiac sales literature stressed some additional suspension advances. A new stabilizer bar was standard on all models and a rear stabilizer was found on Formula 400 and Trans Am models. Also, the use of twin, bucket-type rear seats allowed a higher drive shaft tunnel, so the entire body could actually be mounted lower, for a lower look and center of gravity.

Powerplant revisions were limited mainly to low-priced models. A Chevrolet-built conventional

Completely restyled base Firebird used a Chevrolet-built OHV six-cylinder engine as its standard powerplant. Optional V-8s were available. These were midyear models, usually referred to as 1970 1/2 Firebirds. *Pontiac Motor Division*

six replaced the OHC six. The 350 V-8 came only with two-barrel carburetion and was standard in Esprit and optional in the base Firebird. The 350 had new, lower-compression heads and 10 fewer horsepower than the 1969 small-block. In addition, the 400-ci big-block was offered with two-barrel carburetion and lower-compression heads.

Firebirds were still available with standard or Custom interior trim options. No longer offered was a bench seat, power front seat, or folding-back

A new entry for the Firebird series was the slightly upscale Esprit. Its standard equipment included Custom interior, twin outside rearview mirrors, and extra exterior trim moldings. Such features characterized every Esprit through the 1982 model. *Crestline Publishing*

rear seat. Standard vinyl trims came in five colors. Custom trims, with knit vinyl seat inserts, came in seven colors. There were also two additional combinations with cloth seat inserts. Pontiac provided 15 exterior finish colors. Vinyl Cordova tops came in five different shades.

Each of the four Firebird models had its own assortment of standard features. Many options and accessories were also available, at extra cost—some for all Firebirds and some for specific models only when teamed with other extras.

For the most part, the standard equipment included on each of the four 1970 1/2 models was the same standard equipment that was part of that model for the following 12 years. Rather than repeat this list in the chapters covering 1971–82 Firebirds, the entire, detailed list is given only once, in this chapter. Then, as the later models are discussed, any additions, subtractions, and refinements are noted.

In 1970, the standard equipment for base Firebirds ($2,875) included all GM standard safety features, plus: the Endura front bumper; an all-vinyl interior; front bucket seats; twin rear bucket-type seats; front stabilizer bar; woodgrained vinyl instrument panel; manual front disc and rear drum brakes; 155-horsepower, 250-ci, six-cylinder engine; carpeting; outside rearview mirror; side-marker lights; E78-14 blackwall tires; six-inch-wide wheel rims; and door storage pockets.

Another new model-option was the Formula 400, which basically replaced the 400. Immediate recognition was possible because of the special fiberglass hood with low-riser scoops. Model name also appeared on front fender below Firebird name. White-letter tires became standard on this model as well as the Trans Am. *Pontiac Motor Division*

The Chevy overhead valve (OHV) inline six, although less exotic and popular, was a solid powerplant with 3.87 and 3.53 inch bore and stroke measurements, respectively. Its displacement matched the Pontiac 4.1 liter OHC six, but horsepower dropped to 155 at 4,200 rpm. Optional for the first time in base Firebirds was a V-8. It displaced 350 cubic inches, had 8.8:1 compression, and developed 255 horsepower at 4,600 rpm. Cars with this two-barrel powerplant had red 350 callouts placed under the Firebird lettering on their front fenders.

Promoted as a luxury car, the Esprit used the two-barrel 350 as standard equipment. Its other features were everything found on base Firebirds, plus: a 15/16-inch front stabilizer bar; knit-vinyl (Custom) upholstery; upper level ventilation system; vinyl-clad deluxe steering wheel; dual color-keyed sport outside rearview mirrors; Esprit exterior chrome moldings and nameplates; concealed windshield wipers; three-speed manual (synchromesh) transmission with floor mounted shifter; E78-14 blackwall tires with six-inch-wide rims; fitted trunk floor mats; wheel trim rings; roof drip moldings; and DÈcor trim.

Like other models, the Esprit had its own base retail price ($3,241). Esprit scripts decorated the roof sail panels, and the flush door handles on the exterior had body-color vinyl inserts.

PMD described the Formula 400 ($3,370) as its "road car." Its standard equipment list included all standard GM safety features, plus: the 400-ci, 330-horsepower, four-barrel V-8; heavy duty floor-mounted Hurst shifter; three-speed manual transmission; 7/8-inch front and 5/8-inch rear stabilizer bars; special high-rate springs; special wind-up axle controls; F70-14 bias-belted blackwall tires on seven-inch rims; manual front disc brakes; rear drum brakes; carpets; all-vinyl upholstery; front bucket seats; twin rear bucket-type seats; dual outside rearview sport-type mirrors; concealed wipers; Formula 400 nameplates; and a deluxe steering wheel.

Two 400-ci engines were available. The base edition had a four-barrel carburetor and 10.25:1 compression, resulting in 350 horsepower at 4,800 rpm. Optional was a Ram Air engine with 10.75:1 compression and 345 horsepower at 5,400 rpm.

For exterior identification, the Formula 400 sported a special fiberglass hood with twin, snorkel-type scoops near the front. Call-outs on the front fender, below the Firebird name, read "Formula 400." Cars with ram induction had Ram Air decals on their hood scoops. Also apparent from the outside were twin sport mirrors, chrome exhaust extensions, and concealed wipers.

PMD had originally intended to call this car the Trans Am and rename its top model the Sebring, but Plymouth had "dibs" on the Sebring name, so Pontiac adjusted its plans accordingly.

Trans Am took on more of the sedan racer image with front spoiler, rear wheel flares, and dual sport outside rearview mirrors. There was also a shaker hood scoop—a muscle car item. White-letter tires and bird decal on nose were also introduced. Restyled body was often compared to a Maserati. *Gerry Burger*

Trans Ams became available in blue as well as white. Note the new striping treatment and bird decal. Design of such graphics varied in relation to body color, but the same colors—white, blue, and black—were used in both cases. 1970 was the first year for single headlamps. *Larry DeLay*

With the new body, Pontiac's Trans Am was a knockout. It was a real threat to the Corvette in America's sports car market, as well as worthy competition for foreign exotic cars costing up to five times as much.

That was really something for a car that was basically a Firebird with lots of extras bolted on or bolted in. Among them were the front air dam, rear wheel-opening air dam, rear deck lid spoiler, and front fender air extractors.

Other standard Trans Am features included F60-15 white-letter tires; variable ratio power steering; power brakes (front discs, rear drums); Rally II wheels (less trim rings); dual horns; dual body-color outside rearview mirrors (left-hand remote controlled); concealed wipers; Safe-T-Track differential; and a handling package consisting of tighter-control shocks and heavier front and rear stabilizer bars.

The standard Firebird interior was standard in Trans Ams with some additions. The first was a 14-inch Formula steering wheel with two holes poked into each of the three spokes and a gear-sprocket-shaped hub. The second was a special instrument panel insert. Original sales forms called for a woodgrain insert, similar to the 1969 style. However, the insert used on production models was aluminum with simulated engine turning.

A third distinction of 1970 1/2 Trans Am was the placement of Firebird emblems on the interior. They appeared on the steering wheel hub and in the upper left-hand corner of the aluminum instrument panel insert. In addition, the Trans Am instrumentation package included a Rally gauge cluster with a clock and a tachometer that was oriented to indicate redline at the twelve o'clock position the way many race cars are often set up.

Pontiac sales literature showed that a 400-ci Ram Air V-8 was standard. The catalog noted four-barrel carburetion, 10.5:1 compression and 430 foot-pounds of torque at 3,400 rpm. This was the Ram Air III motor, which most sources list at 335 horsepower.

There was also a 345 horsepower Ram Air IV option available. Ordering it required the use of a Special Equipment Order Form, so it was not an RPO (Regular production option). This powerplant had 10.75:1 heads and a higher redline: 5,400 rpm. It also developed 430 foot-pounds of torque, but at 3,700 rpm.

Rally II wheels and white-letter tires were standard on 1970 1/2 Trans Ams. The front fender air extractor scoops were positioned higher and totally redesigned. Trans Am name decal moved rearward, behind wheel cutout, and stayed below main bodyside feature line. *Larry DeLay*

Firebirds with standard interior (this Trans Am included) had bright metal door handles on cars with the Custom interior option and a body-color insert was applied to the door handle. Since the Custom interior was standard on Esprits, they also had vinyl door inserts as regular equipment. *Larry DeLay*

These were basically the same engines as offered in 1969, except that they could no longer be ordered with a three-speed manual transmission. Other equipment that Trans Am buyers received for their $4,305 included: a four-speed manual transmission; dual exhausts with chrome extensions; chrome engine parts; and a special air cleaner designed for cold-air intake through a special shaker hood with a rear facing air scoop.

If the 1970 1/2 Firebirds were improved in appearance, they were also improved in terms of service reliability. Owner surveys taken in 1971 showed the sixes to be excellent cars requiring fewer than the average number of repairs. The V-8s required about average servicing overall, but needed fewer than-average clutch and brake repairs. For both types of engines, this was a vast improvement over comparable 1967–69 Firebirds.

Owner surveys taken in 1974 showed a few additional problems with 1970 1/2 V-8 models. They were in the areas of exteriors, body integrity, cooling systems, and automatic transmissions. However, the overall service record for these cars still required only the average number of repairs.

The later surveys did not cover six-cylinder Firebirds, because their production was so limited that insufficient data on repairs was available. It can probably be assumed that body deterioration (mainly rust in the lower front fender corners and rockers) was also a problem with sixes. But it is unlikely that the motor had any serious faults, as the same engine in Chevrolets required far less than the average amount of service.

Pontiac serial numbers were still in the same places and used the same basic system. Each of the four Firebirds had a distinct model number (see production chart), coded as the first five characters of the VIN. The sixth character was a zero for 1970 and the seventh was an N for the Norwood, Ohio, factory. The next six digits were numbers indicating a car's order in the production sequence. Numbers 100001 through 600000 were set aside as V-8 numbers, and sixes had numbers from 600001 up. Only 3,134 sixes were built.

As usual, there were alphabetical codes stamped on the engine block (in the same places), which revealed the type of engine and transmission. Engine codes used for 1970 1/2 Firebirds were as follows:

Entire Firebird instrument panel was redone for 1970 1/2 and (this) Trans Am configuration looked particularly nice with simulated engine-turned inserts. The front console was $58.98 extra. Power windows could only be ordered when teamed with the console, where the control buttons were located. Radio options were limited to AM push-button or AM/FM push-button. A stereo tape deck could be added to any car with a radio for $133.76 extra. *Larry DeLay*

Chrome dress-up parts were standard on Trans Am's Ram Air III or Ram Air IV engines. Shaker hood scoop sat atop Rochester Quadrajet carburetor. Seal under scoop was designed to keep water out. *Larry DeLay*

Displ.	Engine	Carburetor	HP	Transmission	Code	Compression
250 cid	L-head six	1-bbl	175	manual	ZB	8.5:1
				automatic	ZG	
350 cid	350 V-8	2-bbl	255	manual	WU	8.8:1
				automatic	YU	
400 cid	400 V-8		265	manual	—	8.8:1
				automatic	XX	
	400 V-8	4-bbl	330	manual	WT	10.25:1
				automatic	YS	
	400 V-8 (Ram Air III)		335	manual	WS	10.5:1
				automatic	YZ	
	400 V-8 (Ram Air IV)		345	manual	WH	10.75:1
				automatic	XN	

The 1970 1/2 models had a shorter-than-usual run. They went on sale around March 1, 1970, and were built only through August. This may be one explanation for a 40,000 unit production decline, but not the only one—there was also a general decline in the popularity of pony-type cars at this time. Firebird sales didn't go much higher in 1971, and the figures for 1972 and 1973 were even lower.

Getting back to the 1970 1/2 models, the

Here's a good view of the Trans Am's integrated spoiler. Dual exhausts with chrome extensions were standard. Trans Am decal on spoiler seemed to conflict with Pontiac nameplate on rear panel.

production figure picture looked like this:

Name	VIN	Manual	Automatic	Total
Base Firebird Six	22387			3,134
Base Firebird V-8	22387	2,899	15,975	15,740
Esprit	22487	2,104	16,857	18,961
Formula 400	22687	2,777	4,931	7,708
Trans Am Ram Air III	22887	1,769	1,339	3,108
Trans Am Ram Air IV	22887	29	59	88
Total		9,578	39,161	48,739

Strato-bucket seats were used with the extra-cost custom interior. They came in blue, brown, red, saddle, green, sandalwood, and black. A pebble-grain vinyl was used on seat borders. Pleated inserts were of a "breathable" woven vinyl. Note custom door panels. *Jack Miller*

The good looks and better reliability of 1970 1/2 Firebirds haven't had a drastic effect on their current desirability as collector cars. The four models range between poor and good on my investment rating scale.

Six-cylinder models, although relatively scarce, aren't nearly as interesting nor desirable as the earlier OHC sixes. You may want one for economical, everyday use or for your child who is going to college, but that's about it.

If you are still determined to invest in 1970 1/2 Firebirds and want one that's going to appreciate faster than average, your only good move is to get a hold of a Formula or Trans Am with the rare Ram Air IV option. Such cars—especially with a four-speed transmission—are hard to find and probably deserve a four-star (very good) investment rating.

As a "sleeper" type of investment, my personal pick would be the Formula 400; one with as many original options as you can find. This model, so far, has remained in the shadow of the Trans Am, but its day in the spotlight could come soon, if general interest in Firebirds continues to grow.

Standard Firebird interior and door panel treatment. This interior was regular equipment in all models except Esprit. It came in metallic blue, saddle, green, sandalwood, and black. Carpeting was standard.

Custom interior was also available in two cloth and vinyl combinations: sandalwood and black. Formula steering wheel is also shown here. The pleated insert material was called Kinross Pattern Cloth. *Jack Miller*

1970 Firebird Options and Accessories

Sales Code and UPC	Description	Retail Price ($)
591	Speedometer gear adapter	11.59
731**	H-D air cleaner (exc. Trans Am; n.a. w/348)	9.48
582	Custom air conditioner	375.99
368	Performance axle	10.53
692	H-D battery	4.21
451	Custom seat belts (front & rear and front shoulder)	12.64
452	Custom seat belts (front & rear and shoulder)	38.97
492	Elec. Clock (exc. Trans Am; n.a. w/652)	15.80
488	Rally cluster w/clock (exc. Trans Am; n.a. w/341, 484, 652)	47.39
484	Rally cluster w/clock (std. w/Trans Am; n.a. w/341, 484, 652)	94.79
494	Front console (na in six w/three-speed trans.)	58.98
402	Spare tire cover	5.27
481	Cruise control	57.93
541	Rear window defogger	26.33
534	Elec. Rear window defroster (V-8 only)	52.66
361	Safe-T-Track differential (std. w/Trans Am; req. 37R axle)	42.13
343	350-ci V-8 2V (std. w/22487; n.a. w/351, 358)	110.59
348	400-ci Ram Air V-8 2V (Esprit 22487 w/Code 350)	52.66
	(std. w/Trans Am; n.a. w/352, 359, 754)	168.51
711	Evaporative emission control (req. in Calif. Cars)	36.86
531	Soft-Ray tinted glass, all windows	32.65
532	Soft-Ray tinted glass, windshield only	26.33
684	Door edge guards	6.32
674	Dual horns (22387 base Firebird; std. in others)	4.21
661	Convenience lamps	11.59
521	Front floor mats, pair	6.85
522	Rear floor mats, pair	6.32
524	Fitted trunk floor mats (std. w/22487; incl. W/731)	8.43
434	Dual body-color sport-type OSRV mirrors (22387)	26.33
441	LH visor vanity mirror	2.11
431	Decor moldings (22387, 22687 only; std. on others)	47.49
741	Roof drop moldings (std. w/22487)	31.60
634	Rear lamp monitor	26.33
754	Mountain perf. axles	
	22387, 22487 w/o 582	16.85
	343 w/352, 359; 22387, 22487 w/582	2.11
502	Wonder Touch power brakes (std. w/22887)	42.13
554	Remote-control deck lid release	14.74
552	Power door locks & seatback locks (na w/734)	68.46
734	Power door locks only	45.29
501	Variable-ratio power steering (std. w/22887)	105.32
551	Power side windows (all w/code 494)	105.32
401	AM push-button radio	61.09
402	AM/FM push-button radio	133.76
404	AM/FM stereo radio	239.08
411	Rear speaker	15.80
412	Stereo tape player (reg. w/radio; n.a. w/411)	133.76
SVT	Cordova vinyl top	
	22387, 22687 w/o 431	89.52
	22387, 22687 w/431 and 22487	73.72
461	Deluxe steering wheel (22387; std. w/22487, 22687)	15.80
464	Formula steering wheel	
	w/501 only; in 22487; 22687	42.13
	w/501 only; in 22387; n/c in 22887	57.93
504	Tilt steering wheel (reg. w/601; n.a. w/col. shift)	45.19
351	Turbo Hydra-matic trans.	
	22487, 22687; n.a. w/341, 343	227.04
	22887	n/c
352	2-speed auto. Trans.	
	22387 w/6-cyl. 341	163.68
	22387, 22587 w/343 V-8 350 ci 2V engine	174.24
354	4-speed manual trans. w/floor shift (req. V-8; n.a. w/754)	195.36
355	H-D 3-speed man. trans. w/floor shift (w/o 341; std. w/22687)	84.26
358	Close-ratio 4-speed trans. w/floor shift (22687 w/37R axle; 22887 w/37R axle n/c; req. w/361)	195.36
359	Turbo Hydra-matic	
	22387, 22487 w/343 engine	205.92
	22387 w/341 6-cyl. engine	195.36
473	Wire wheel discs	
	22487	52.66
	22387, 22687	73.72
474	Rally II rims	
	22487	63.19
	22387, 22687 (incl. 471; std. w/22887)	84.26
471	Wheel trim rings (22387, 22687 w/o 454; std. w/22487)	21.06

Sales Code and UPC	Description	Retail Price ($)
432	Recessed windshield wipers (22387, 22687; std. w/others)	18.96
321	Basic group (all exc. 22887)	103.22
731**	Custom trim group (std. 22487)	77.94
331†	Power assist group	
	22387 w/6-cyl. engine 341	342.81
	22387, 22487 w/V-8 engine 343	353.37
	22487 w/400-ci V-8 engine 346; 22687	374.49
652††	Warning lamp (exc. 22887; n.a. w/484 or 488)	36.86
TGR	F78-14 whitewall tires (22387, 22487 w/341, 343, 346)	43.18
THR	E78-14 whitewall (22387, 22487 w/341, 343, 346)	26.33
TMF	F70-14 blackwall tires (22387, 22487 w/341, 343, 346)	35.81
TML	F70-14 raised-letter tires	
	22387, 22487 w/341, 343, 346)	64.25
	22687	28.44
TNL	F60-15 raised-letter tires (22687; incl. 15-inch Rally II)	146.39
704	Space-Saver spare tire	
	Exc. 22887; std. w/474 or TNL	15.80
	22887	n/c

*Includes AM radio, electric clock, visor vanity mirror, outside rearview mirror, remote-control mirror, and heavy-duty air cleaner.

**Includes instrument-panel assist group, bright roof rail side interior moldings, custom front seat, door and quarter panel interior trim, and molded trunk mat.

†Includes Turbo Hydra-matic transmission, power steering, and power brakes.

††Includes warning light for low fuel, low washer fluid, seat belt, and headlamp.

Chapter 5

1971

★	Firebird Six
★	Firebird V-8
★	Esprit
★★	Esprit 400
★★	Formula 350
★★★	Formula 455
★★★★	Formula 455 HO
★★★★	Trans Am

All 1971 Firebirds were carried over from the 1970 1/2 series with minor changes: seat construction was revised; new bucket seats were of high-back design; and dummy air vents with horizontal louvers were added to all models except the Trans Am (they were located below the main bodysided feature line and behind the front wheel cutouts).

The base Firebird was $3,047 with a six, or $3,168 with a two-barrel 350 V-8. There were no changes in standard equipment from 1970, but to comply with federal requirements, the front side-marker lights were rewired to blink in unison with turn signals. Again, 15 exterior colors were offered, but a dozen of them were new.

On the inside of the cars, the new Custom interior options were dark saddle, sienna, and jade green, replacing red, saddle, and green. Outside, Cordova vinyl tops were again offered in five colors with dark brown replacing dark gold. New options included a rear console, a cassette tape player (the only one offered in a U.S. car), and a rechargeable glovebox flashlight. Honeycomb-styled wheels—with a spoke pattern resembling a section of a beehive—were introduced as optional equipment for Firebird, Esprit, and Formula models, and were standard equipment for the Trans Am.

1971 Firebird. *Pontiac Motor Division*

Chevrolet produced the OHV six-cylinder engine used in only 2,975 base Firebirds that season. Like all new engines, it came with sealed and preset carburetor mixture screws; a more-sensitive aluminum choke housing; and a self-regulating, integral circuit alternator. Featuring a one-barrel carburetor and 8.5:1 compression, it was rated for 155 horsepower at 4,200 rpm and 235 foot-pounds of torque at 1,600 rpm. The standard gearbox was a column-shifted three-speed manual unit.

GM was in the process of changing to the Society of Automotive Engineers' net horsepower ratings in 1971, and it listed both gross (brake) and net ratings. Under the new system, the six carried a 110 net horsepower figure. The net horsepower rating system was used exclusively beginning in 1972. I'll use the abbreviation nhp to stress the difference from gross brake horsepower (bhp) in this book.

Transmission options for the 1971 Firebird six included column-shift controls and a two-speed automatic that it shared with only small V-8s. The optional 350 V-8, which was also standard for Esprit and Formulas, had 8.8:1 compression. It was rated for 250 gross horsepower at 4,600 rpm and 335 foot-pounds of torque at 2,800 rpm. The second rating for it was 165 net horsepower. The V-8 could be had with the standard three-speed manual transmission (including floor shifter), the four-speed manual gearbox, a two-speed automatic, or a three-speed Turbo Hydra-matic automatic. Cars with this option had a 350 call-out below the Firebird lettering on the front fender.

With a $3,416 base sticker price, the Esprit continued as a luxury touring version of the Firebird. The only change in its standard equipment list was that a Custom cushion steering wheel replaced the former deluxe type. Esprit's base engine was the two-barrel 350 V-8 attached to the regular three-speed manual gearbox with a floor-mounted shifter. Buyers could also order a heavy-duty three-speed manual transmission, two-speed automatic, four-speed manual transmission, or Turbo Hydra-matic automatic. An available engine option was a 400-ci V-8 with a two-barrel carburetor and 8.2:1 compression. It was rated for 265 gross horsepower at 4,400 rpm and 180 net horses. The torque figure was a 400 foot-pounds at 2,400 rpm. This powerplant was available only with the Turbo Hydra-matic transmission.

For 1971, the Formula's factory price increased to $3,440. About the only change in standard equipment was the new Custom cushion steering wheel. New engine options created a Formula 455 model to join the base Formula 350 and the Formula 400. More engines were available in this model than in any other 1971 Firebird. Each Formula carried appropriate front fender call-outs to announce its engine displacement.

There were a number of new extras offered for all three Formulas. They included a rear deck lid spoiler and a special version of the optional Y96 Ride & Handling package that included the honeycomb wheels.

Close-up shows simulated louvers found on all 1971 models except Trans Am. These were a one-year-only feature. *Pontiac Motor Division*

A new 1971 option were polycast honeycomb wheels. Designer Bill Porter was responsible for the design.

The base Formula powerplant was the two-barrel 350 linked up to the floor-shifted heavy-duty, three-speed, manual gearbox. Its transmission options included the two-speed automatic, a three-speed Turbo Hydra-matic, a four-speed manual, and a close-ratio four-speed manual.

Next came one of two engine options that were exclusive to the Formula series: the four-barrel version of the 400 V-8 with 8.2:1 compression and 300 gross horsepower at 4,800 rpm. This engine's net output was 250 horsepower and its torque rating was 400 foot-pounds at 3,600 rpm. Gearbox options included all those listed for the Formula 350, except the two-speed automatic.

Two 455-ci powerplants were also available in the Formula. The more powerful version was the 455 HO (described later), which was standard in the Trans Am. The less-powerful version also used a four-barrel carb with a more modest 8.2:1 compression ratio. It produced 325 gross horsepower at 4,400 rpm and was also rated for 255 net horsepower. The torque figure stood at 455 foot-pounds at 3,200 rpm. Transmission linkup with this engine was limited to the Trans Am-type three-speed Turbo Hydra-matic, which was beefed up from standard specifications to handle the extra horsepower and torque. The 455 HO, on the other hand, could be had (in Formulas) with the heavy-duty three- or four-speed manual gearboxes.

With a base retail price of $4,464, the Trans Am was the least changed of all 1971 Firebirds on the outside, unless you include the wheels. Standard equipment was switched from Rally II rims without trim rings to the honeycomb wheels.

Under the hood was another change, substitution of the new 455-ci High Output engine for 1970's 400-ci Ram Air III. The 455 HO had 8.4:1 compression heads and pistons plus the functional cold-air induction system. Its gross horsepower rating was 335 at 4,800 rpm with 480 foot-pounds of torque at 3,600 rpm. Its net horsepower rating was 305.

As in the past, the Trans Am was offered only with a limited number of exterior finish colors. Cars done in cameo white had blue stripes on a black base. Also available was a Lucerne blue body finish with white stripes on a black base. Two minor changes for the year were a smaller (17-gallon) fuel tank and the absence of a chrome engine dress-up kit on the standard equipment list.

Like the 1970 1/2 Firebirds, the 1971 models had several components that were worse than average in terms of reliability. *Consumer Reports 1973 Buyer's Guide* said that the V-8 models had a noticeable number of problems in three main areas: body integrity, manual transmission, and

The Formula also used the new front fender louvers and high-back bucket seats. Formula identification on front fender below louvered trim identified series name and engine displacement. *Pontiac Motor Division*

shock absorbers. All other systems on the V-8 models were listed as requiring the average number of repairs. However, the problem areas brought the overall opinion into the "worse-than-average" category.

Pontiac eventually cured the shortcomings of the Firebird gearbox and suspension, but body integrity complaints were commonly heard well into the 1980s. With the second-generation models, the cause seemed to be that their separate subframe construction compromised structural rigidity of the body.

Consumer Reports' later surveys didn't include the six-cylinder cars, most likely because there weren't enough built to provide sufficient repair data. However, the same six-cylinder engine in Chevrolets was rated very highly, so it's likely that the Firebird sixes had a slightly higher reliability rating than the V-8s.

Despite its good service record, there was a factory recall on the 1971 six-cylinder engine. Dealers were advised to check the carburetor attachment clip (from throttle rod to throttle lever) to ensure it had been correctly installed. This was a minor problem that was easily repaired on most of the cars when they were new.

Magazine road tests on 1971 Firebirds are hard to come by. The performance era was dying and it appears that auto makers didn't want to rock the boat by having their hottest models nationally publicized. As for the magazines, their editorial content followed the trend to smaller and the more economical automobiles.

The English magazine *Motor* was one exception. Its editors gave a very complete, full-range test to the Formula 400 Firebird in July 1971. The car had the standard setup of a four-barrel 400 linked to the three-speed Turbo Hydra-matic transmission. Performance figures showed a 0–60 mile per hour time of 6.5 seconds, and the car did the quarter-mile in 14.8 seconds. Top speed was recorded as 131.3 miles per hour. In 1,870 miles of total use, the Formula 400 had an overall fuel consumption rating of 12.1 miles per gallon, but touring economy at 16 miles per gallon wasn't too bad for that period.

Pontiac continued to use the same 13-character VIN system that was instituted in 1969. Again, the first five characters represented the model number, the sixth identified the year, and the seventh was a letter pinpointing the assembly plant (which was still the Norwood, Ohio, factory in the case of all Firebirds). The last six digits were the sequential production number.

Changes in 1971 Trans Am included high-back bucket seats. Under the hood, only one engine was used: the 335-horsepower 455 HO V-8. Chrome engine dress-up parts were no longer standard equipment.

The rechargeable flashlight option was developed by GM's Delco-Remy Division. This advance publicity photo indicates woodgrained dash inserts may have been planned for reintroduction in 1971. However, in production the simulated engine-turned inserts continued to be used. Photos like this are usually taken around six months to a year before actual production dates.

Sequential numbers set aside for V-8 models were 100001 through 600000, while the sixes ran from 600001 up.

As in the past, alphabetical codes were stamped in the regular locations on the engine blocks to identify the type of engine installation and the original transmission attachment. Engine and transmission codes for the 1971 models can be translated according to the following list:

This is one of only 885 Trans Ams made with stick shift in 1971. Interior was unchanged from 1970, except for seat design.

Model-year production total climbed a little in 1971, hitting a grand total of 53,124 units, versus the previous season's (short run) 48,739. Available breakouts by model, engine, and transmission looked like this:

Name	VIN	Manual	Automatic	Total
Base Firebird Six	22387			2,975
Base Firebird V-8	22387	2,778	20,244	20,047
Esprit	22487	947	19,238	20,185
Formula	22687	1,860	5,942	7,802
Trans Am 455 HO	22887	885	1,231	2,116
Total		6,470	46,655	53,125

Engine	Carburetor	Horsepower	Transmission	Compression	Code
250 ci 6-cyl	1-bbl	155	manual	8.5:1	CAA, ZB
			automatic		CAB, ZG
350 ci WP	2-bbl	255	manual	8.8:1	WR, WU, WN,
			automatic		XR, YU, YN, YP
400 ci		265	manual	8.2:1	WS, EX
			automatic		XX, YX
	4-bbl	300	automatic		YS
455 ci		325	manual		WJ
			automatic		YC
455 ci HO		335	manual	8.4:1	WL, WC
			automatic		YE

As collector cars, the 1971 Firebirds are starting to catch on, and the Trans Am is the most desirable model to have. With a total production run of only 2,116 units, finding a good car can be a chore—although your efforts are practically guaranteed to pay off in the long run.

Classic car auction reporter Jerry Heasley noted, "you can't go wrong with a '71 (Trans Am) from an investment point of view. Just make sure the car you buy is complete."

If you should manage to locate two 1971 Trans Ams with different transmissions, the wiser choice would be the manually-shifted car. Don't worry too much about the fact that gearbox service problems were common; a stick shift is relatively inexpensive and easy to fix. Finding one of the 885 Trans Ams built with this transmission is a little harder. The four-speed will definitely add to the car's collector market value. However, with only 1,231 assemblies, the Turbo Hydra-matic isn't exactly what you would call a common setup.

If Trans Ams seem a little out of your price range, a good bet for a "sleeper" value would be the 1971 Esprit with stick shift and the optional 400-ci engine. This combination must be rather low in supply, as only 947 Esprits had manual gearboxes to begin with. When you throw in the engine option, you are looking at added desirability and rarity.

The thing to keep in mind is that other collectors usually aren't after Esprits. The Trans Am is the glamour car, and Formulas are just starting to catch on. So you may be able to find a real bargain in the Esprit category, by just knowing what to look for.

While you're checking the classified ads for Trans Ams and stick-shift Esprits, it won't hurt to keep your eyes peeled for Formula 455s, especially examples with a four-speed manual transmission.

The production totals don't show specific breakouts for Formulas with different engines, but two factors can give you an idea of how many were produced: One, total production for all three models—Formula 350, Formula 400, and Formula 455—came to just 7,802 cars; and two, the various Formula options incorporated four different engines and five different transmissions (certain restrictions limited the total number of possible combinations to about 15).

Now suppose PMD built an equal number of each possible combination. Dividing 7,802 production units by 15 possible power-team combi-

The 1971 Trans Am 455 HO engine was identified by a small decal on the shaker scoop. This decal was in body color, and contrasted with the scoop finish. Scoop was blue on white cars or white on blue cars. Note the look of chrome engine kit.

Custom interior for the 1971 Firebird, actually a 455 HO Trans Am. Seat design changed from 1970.

nations, you get an average of 560 cars for each. In real life, the base 350 with Turbo Hydra-matic probably drew the majority of orders, which means the number of higher-dollar variations— say, a Formula 455 HO with four-speed manual transmission—must have been produced and sold in very limited numbers. That's something else to keep in mind when you're out car hunting.

1971 Firebird Options and Accessories

Sales Code and UPC	Description	Retail Price ($)	Sales Code and UPC	Description	Retail Price ($)
321/Y88*	Basic option group		SVT/C08	Cordova top	
	2387 6-cyl.	448.17		2387, 2687	89.52
	2387 V-8	458.73		2487	73.72
	2487 V-8	432.40	421/a90	Remote deck lid release	14.74
	2687 w/350 V-8	471.36	541/C50	Rear window defogger (na w/C49)	31.60
	2687 w/optional V-8s	492.48	543/C49	Elec. rear window defroster	
321/Y88*	Basic group; AM radio only (2887)	66.35		(na w/34A or C50)	63.19
331/Y96**	Handling package (2687; incl. P05)	205.37	554/AU3	Power door locks	45.29
34DX/L30	350 V-8 2V (2387)	121.12	691/WU1	Self-charging flashlight	12.64
34G/L65	400 V-8 2V (2487)	52.66	718/W63	Rally gauges and clock	
34L/L78	400 V-8 4V (2687)	100.05		(na w/34A, 652; incl. U35)	47.39
34P/L75	455 V-8 4V (2687 only)	157.98	714/U30	Rally gauges, clock, instrument panel	
34U/LSZ5	455 HO V-8 4V			tach. (na w/34A, 718, 7822, 652)	94.79
	(2687 only; std. w/2887)	236.97	532/A02	Soft-Ray glass, windshield only	30.54
35K/M38	3-speed turbo Hydra-matic trans.		531/A01	Soft-Ray glass, all windows	37.92
	w/6-cyl. 34A engine	175.26	492/B93	Door edge guards	6.32
	w/350 V-8	183.96	734/V32	Rear bumper guards	15.80
35L/M40	3-speed turbo Hydra-matic trans.		601/WU3	Hood air inlet	
	(n/c w/2887)	201.48		(avail. w/455 HO only on 2687)	84.26
35J/M53	2-speed auto. trans.		681/U05	Dual horns	
	w/6-cyl. 34A engine	148.92		(2387 only; std. w/all others)	4.21
	w/350 V-8; n.a. w/2687	190.00	652/W74	Warning/clock lamps	
35A/Std.	3-speed manual trans.; column			(na w/714, 718, 722)	42.13
	(2387 6 cyl. only)	n/c	664/Y92	Convenience lamps	11.59
35B/M12	3-speed man. trans.; floor shift		521/B32	Front floor mats only	7.37
	(w/34A, 34D)	10.53	522/B33	Rear floor mats only	6.32
35C/M13	H-D 3-speed man. trans.; floor shift		524/B42	Rear compartment mat (std. w/2487)	8.43
	(na w/34A, 34G)	84.26	441/D34	RH visor vanity mirror	3.16
35E/M20	4-speed man. trans. (na w/34A, 2887)	205.97	434/D35	Dual sport OSRV mirror	
35G/M22	Close-ratio 4-speed man. trans.			(LH remote; 2387 only)	26.33
	(na w/34A, 34D; std. w/2887)	237.65	484/B85	Belt reveal moldings (std. w/2487)	21.06
361/G80	Safe-T-Track rear axle (std. w/2887)	46.34	491/B96	Front and rear wheel opening	
368/G90-2	Performance axle	10.53		moldings (std. w/2487; n.a. w/2887)	15.80
422/K45	Dual-stage air cleaner (na w/2887)	9.48	481/B80	Roof drip moldings	
582/C60	Custom A/C (na w/34A)	407.59		(incl. w/SVT; std. w/2487)	15.80
654/TP1	Delco X battery (w/455 ci V-8 only)	26.33	494/B84	Vinyl body side moldings	
692/UA1	H-D battery (na w/582)	10.53		(black; n.a. w/2887)	31.60
502/J50	Power front brakes (std. w/2887)	47.39	704/WT5	Mountain performance option	
722/U35	Elec. clock (incl. w/W74, U30, W63)	15.80		w/C60	10.53
431/D55	Front console (w/floor shift only)	58.98		w/o C60	31.60
424/D58	Rear console	26.33	701/V01	H-D radiator	21.06

Sales Code and UPC	Description	Retail Price ($)
401/U63	AM radio w/windshield antenna	66.35
403/U69	AM/FM radio & windshield antenna	139.02
405/U58	AM/FM stereo radio & windshield antenna (na w/U80)	238.08
451/AK1	Custom safety belts	15.80
684/N65	Space-Saver spare	15.80
411/U80	Rear seat speaker (na w/U58, U55, U57)	18.96
572/D80	Deck lid spoiler (std. w/2887; avail. w/2687)	32.65
501/N41	Variable-ratio power steering (std. w/2887)	115.85
461/N30	Custom cushion steering wheel (std. w/2487, 2687; n.a. w/2887)	15.80
464/NK3	Formula steering wheel	
	2487, 2687	42.13
	2387; std. w/2887	57.93
504/N33	Tilt steering wheel (na w/manual steering & column shift)	45.29
414/U55	Cassette tape player (na w/U80, U57)	133.76
412/U57	Stereo 8-track player (na w/411, 414)	133.76
471/P06	Chrome wheel trim rings (std. w/2487)	26.33
731/W54†	Custom trim option (std. w/2487; n.a. w/2387)	78.99
472/P02††	Custom wheel covers	
	2387, 2687	31.60
	2487	5.27
478/P05	Honeycomb wheels	
	2387, 2687	126.38
	2487	100.05
	2887	36.86

Sales Code and UPC	Description	Retail Price ($)
474/N98	Rally II wheels	
	2387, 2687; n/c w/2887	89.52
	2487	63.19
NL/PM7	Rally II wheels and F60-15 W/L tires (2687; std. w/2887)	162.19
NL/PM7	Honeycomb wheels and F60-15 W/L Tires (2687; std. w/2887)	199.05
551/A31	Power windows (req. w/D55)	115.85
432/C24	Concealed wipers (std. w/2487, 2887)	18.96
HR/PL3	E78-14 whitewall tires (2387, 2487)	28.44
GR/PX6	F78-14 whitewall tires (2387, 2487)	45.29
ML/PL4	F70-14 white-letter tires	
	2387, 2487	76.88
	2687	41.07
MF/PY6	F70-14 blackwall tires (2387, 2487)	35.81

*Includes AM radio with windshield antenna, THM trans., std. size whitewall tires, custom wheel trim rings, and power steering (std. w/2887).

**Includes honeycomb wheels; F60x15 white-letter tires; heavy front and rear stabilizer bars and heavy-duty rear springs.

†Includes assist straps on door and above glovebox, perforated headliner, formed vinyl trunk mat, and body-color outside door handle inserts.

††Wire wheel covers were no longer available as a factory-installed option. They were available, however, as dealer-installed accessories.

Chapter 6

1972

★	Firebird Six
★	Firebird V-8
★	Esprit
★★	Esprit 400
★★	Formula 350
★★	Formula 400
★★★	Formula 455 HO
★★★★	Trans Am 455 HO

In 1972, General Motors considered dropping the Firebird car line. The reasons included the government's crackdown on high-performance cars and the excessively high rates being charged to insure such vehicles. These factors, combined with the American car buyers' interest in small, economy cars, kept Firebird sales extremely low.

Because of its cloudy future, the Firebird series received very few appearance or technical changes and was virtually ignored in terms of advertising and promotion. For example, Pontiac did not release any separate sales literature for Firebirds, with the exception of a color postcard. Four pages of the company's full-line catalog presented the four basic models.

Styling wise, a new, elongated honeycomb mesh appeared in the Firebird grille. It was designed to go with the honeycomb wheels,

One change in the Esprit and Firebird hardtop coupes was a new honeycomb mesh for the grille insert.

introduced as a 1971 option. These wheels came in two sizes: 14 inches for Firebirds, Esprits, and Formulas; and 15 inches for Trans Ams.

The base Firebird was priced at $2,837. Gone from its front fenders were the dummy air vents of 1971. In their place were Firebird lettering and emblems. Cars with V-8s had a 350 call-out badge below this trim. The rocker panels carried fine-line chrome moldings, and small hubcaps were used.

Standard equipment was unchanged from 1971. There were 10 new color choices, although 10 previous offerings were deleted. This kept the total of color selections at 15 for the sixth year in a row. Changes in interior choices cut the total number from 14 to 12. Standard Madrid-pattern Morrokide trims came in ivory, saddle, green, and black.

All Firebirds could again be had with Custom trim option as either standard or extra cost features. The all-Morrokide (vinyl) styles featured perforated Roulet Morrokide in blue, ivory, saddle, green, covert beige or black. The Potomac-pattern cloth and Madrid Morrokide combinations came in covert beige, or black.

There were a couple of significant changes in powerplants to satisfy new emissions regulations and make up for the resulting horsepower reductions. Smaller choke openings and new, evaporative canister purge valves were used to cut pollution. Larger exhaust valves, new carburetor circuitry, and different spark plugs were used in some engines.

The base Firebirds featured the same six-cylinder engine used in 1971, but only SAE net horsepower and torque ratings were listed in the specifications. The six was rated for 110 net horsepower at 3,800 rpm and its net torque was again 185 foot-pounds at 1,600 rpm. Late in the season, the two-speed automatic transmission was discontinued. Thereafter, the three-speed Turbo Hydra-matic was used in all cars.

A 350-ci, two-barrel V-8 with single exhaust was again optional. It was also coded the same as before and featured 8.0:1 compression. This 1972 motor was slightly down-rated with 160 net horsepower at 4,400 rpm and 270 foot-pounds of net torque at 2,000 rpm.

For 1972, the Esprit was priced at $3,193. It had a wider rocker panel molding, Esprit sail panel scripts, dual color-keyed outside rearview mirrors, a stainless steel roof gutter, hidden

Pontiac described the Formulas as "not-as-smooth-riding" as regular Firebirds. Polycast wheels were an extra; they cost the least on Esprit and slightly more on base Firebird or Formula. This Formula 455 has colored door handle inserts, indicating Custom interior option. *Pontiac Motor Division*

wipers, color-keyed outside rearview mirrors, and color-keyed vinyl door handle inserts to set it apart from lower-priced models. The Custom cloth and Morrokide interior was standard.

A slightly more powerful two-barrel 350 V-8 with dual exhausts was under the Esprit's hood. It produced 175 net horsepower at 4,400 rpm and 275 foot-pounds of torque at 2,000 rpm. A 400-ci two-barrel V-8 with dual exhausts was under the Esprit's hood. It produced 175 net horsepower at 4,400 rpm and 275 foot-pounds of torque at 2,000 rpm. A 400-ci two-barrel V-8 with 175 net horsepower at 4,000 rpm was optional, had 310 foot-pounds of net torque at

2,400 rpm, and could be ordered only with a second type of Turbo Hydra-matic transmission (see chart below):

1972 Turbo

Hydra-matic	1st gear	2nd gear	3rd gear	Overall
Code P	2.52	1.52	1.00	1.92
Code O	2.48	1.48	1.00	2.08

Formula Firebirds included three editions: 350, 400, and 455 HO. The 350 engine was considered optional and featured dual-exhaust. Four-barrel carburetion was standard on the Formula 400 model, also with dual exhausts. This engine had an 8.2:1 compression ratio, 250 net horsepower at 4,400 rpm, and 325 foot-pounds of torque at 3,200 rpm. It came with the four-speed, close-ratio four-speed or Turbo Hydra-matic transmission. The Trans Am's 455 HO V-8 was the top engine option for Formulas.

Pricing for the three Formulas started at $3,221. Externally, a special dual-scoop fiberglass hood characterized the line. The model name appeared on the front fenders with engine call-outs preceding it. Other equipment features were carried over from 1971. There were some changes in the way options were merchandised (for example, honeycomb wheels were no longer required with the Ride & Handling package).

Dual sport outside rearview mirrors were standard on Trans Ams after 1970. The driver's-side mirror was always remote control. *Dave Poltzer*

Body-color door handle insert indicates use of the Custom trim option, which was again the standard fare for Esprits.

The only engine for 1972 Trans Ams was the 455 HO, which produced 300 horsepower at 4,000 rpm. The car shown here has the unitized ignition system new for the year. *Dave Poltzer*

Also, little was changed for the 1972 Trans Am. Rally II wheels with trim rings were made standard equipment, along with unitized ignition, and the close-ratio four-speed. The chrome engine dress-up kit was demoted to an option.

The 455 HO engine was again the Trans Am powerplant. On paper it looked slightly downrated, with 300 net horsepower at 4,000 rpm. Net torque was 415 foot-pounds at 3,200 rpm. An 8.4:1 compression ratio was standard.

Although the new SAE horsepower rating system made the Trans Am seem emasculated, it wasn't. This was driven home in a March 1972 *Car Craft* magazine road test that showed a quarter-mile time of 14.3 seconds with a 98-mile-per-hour terminal speed.

In his book *Supertuning Your Firebird Trans Am,* Joe Oldham speaks highly of this powerplant. "Pontiac continued to have the guts to market a no-holds-barred performance car like the Trans Am with a 455 HO option," wrote Oldham. He also reported testing a car that ran low 14s in stock trim and 13-second quarter-miles when slightly modified with full exhausts.

Color choices for the Trans Am were still limited to cameo white with blue stripes or Lucerne blue with white stripes.

Firebird VIN numbers were again found on the left-hand side of the instrument panel. The coding system was slightly modified, however. A single letter was used to denote the Firebird series, and the fifth character was another single letter revealing the type of engine.

This meant that there were still 13 characters. The first was a 2 for the Pontiac Division. The second designated the series: S on base Firebirds, T on Esprits, U on Formulas, and V on Trans Ams. The third and fourth characters for all Firebirds were 87 for the hardtop coupe body. Fifth came the new alphabetical engine code, and sixth, a 2 for 1972. The seventh characters was an N for the Norwood assembly plant. The last six numbers were the sequential unit number, running from 500001 up. The engine codes for 1972 Firebirds were as follows:

D = 250-ci six
M = 350-ci 2V (single exhaust)
N = 350-ci 2V (dual exhaust)
R = 400-ci 4V (single exhaust)
T = 400-ci 4V (dual exhausts)
X = 455 HO-ci 4V (dual exhausts)

There were also codes stamped directly on the block again. They identified engine and transmission combinations as follows:

This Custom trim door panel pattern was used for several model years. (The interior door lock buttons are aftermarket.) It had molded storage bins and armrest, door pull-handle, and door and manual-window controls. Power window buttons were a variation. *Dave Poltzer*

Formula steering wheel was available on all Firebirds and mid-size Pontiacs. It was standard in the Trans Am and optional for all other models. This 1972 dash shows air conditioner installation.

Horsepower	Displacement	Carburetor
110	250 ci six	IV
160/175	350 ci V-8	2V
175	400 ci V-8	
250	400 ci V-8	4V
300	455 ci HO V-8	

Transmission	Code	Compression
manual	CBG, CBA*	8.5:1
automatic	CBJ, CBC*	
manual	WR	8.0:1
automatic	YR, YV	
manual	YX, ZX*	8.2:1
automatic	na	
manual	WK, WS	
automatic	YS	
manual	WD, WM	8.4:1
automatic	YV, YE	

*California only

The VIN's engine code tells you which power-plant should be in the car. The block code tells you the engine that is in the car, if the block is of the right year, and what transmission it was originally mated with. Codes should match up accordingly. Beginning in 1972, the use of identi-fication stickers traditionally found on air clean-ers was discontinued. No name or displacement figure appeared on the engine.

Total Firebird production fell by 23,173 cars in 1972. PMD's engineering specifications section kept complete records by series and trans-mission type. They showed the following:

Model	Name	VIN	Manual	Automatic	Total
72-23	Firebird	223	1,263	10,738	12,001
72-24	Esprit	224	504	10,911	11,415
72-26	Formula	226	1,082	4,167	5,249
71-27	Trans Am	228	458	828	1,286
	Total		3,307*	26,644	29,951

*Four-speed manual transmission was used in 2,689 cars.

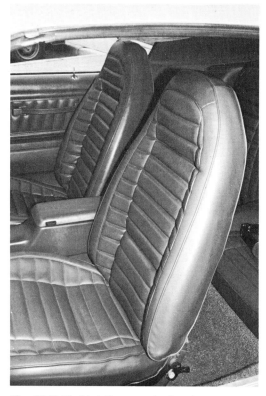

The 1972 Firebird Custom trim interior was available in two variations. One was vinyl with perforated Roulet vinyl insert (as shown here). This option came in blue, ivory, saddle, green, beige, and black. *Dave Poltzer*

Rear passenger compartment of Dave Poltzer's coupe shows perforated-vinyl Custom interior trim and installation of rear package shelf radio speaker. Vinyl pad covers transmission hump between seats.

The standard bucket seats (shown here) were available in Madrid Morrokide for the Firebird, Formula, and Trans Am hardtop coupes only.

Custom bucket seats were available in perforated Roulet Morrokide, Potomac cloth, and Madrid Morrokide (shown here). They were standard on the Esprit hardtop coupe only—available at extra cost on Firebird, Formula, and Trans Am hardtop coupes.

Surveys of 1972 Firebird buyers, compiled in 1974, showed a generally improved service history. Cars of this model year still had some exterior body problems with paint, trim, and (mainly) corrosion. Clutches also seemed to require an above-average number of repairs.

Pontiac, however, solved some other problems that plagued earlier Firebirds in the chassis and electrical system. Exhaust and cooling systems were also made more reliable, perhaps because of the cleaner running engines and lower compression ratios. There was slightly more service required on the 1972 fuel system, indicating the new carburetor circuitry was not fully bug-free. Overall reliability was rated as average, but these were clearly the best Firebirds since the series began, proof of which comes from the fact that there were no factory recalls in 1972.

One thing that may have partly accounted for the Firebird's sluggish sales pace was its gas mileage. Economy was not one of the series' strong points, even with the Chevy built six-cylinder powerplant.

As in the past, the Firebird remained basically an enthusiast's car. It went faster than most 1972 American autos and handled well, too. Ride quality, however, was still tight and firm—not everyone's cup of tea.

For today's investor, the top 1972 models to look for are the Formula with the 455 HO engine and the Trans Am. A blue Trans Ams with a four-speed transmission is considered particularly hard to find; consequently, they are very desirable.

A possible "sleeper" in the collector marketplace is the Esprit with manual transmission. Only 504 of these cars were built with this option and they are probably rare today—rarer than most people realize. They would be most desirable when equipped with the optional 400-ci engine and a four speed.

1972 Firebird Options and Accessories

Sales Code and UPC	Description	Retail Price ($)	Sales Code and UPC	Description	Retail Price ($)
SPS/W51	Solid special paint (w/vinyl top, exc. Trans AM)	$ 97.00	464/NK3	Formula steering wheel base w/501 only	56.00
SPS/W51	Solid special paint (w/o vinyl top; exc. Trans Am)	113.00		Esprit, Formula	41.00
SVT/C08	Cordova vinyl top (Esprit)	72.00	462/N31	Custom sport steering wheel (incl. w/332)	56.00
SVT/C08	Cordova vinyl top (base/Formula)	87.00	471/P06	Custom wheel trim rings (incl. w/474, 332)	26.00
TNL/YH99	Formula Handling package	87.00	472/P02	Custom finned wheel covers	
34H/L30	350 V-8 2V (base)	118.00		base, Formula w/o 332, 474, 478	50.00
34R/L30	350 V-8 2V (Calif.; 35K req. w/base)	118.00		Esprit	24.00
34R/L65	400 V-8 2V (Esprit w/35L only)	51.00	474/N98	Rally II wheels	87.00
34S/L78	400 V-8 4V (Formula)	97.00	476/P01	Deluxe wheel covers	
34X/LS5	455 HO V-8 (Formula; req. w/601, 634; w/35L, G)	231.00		base, Formula, n.a. w/332	31.00
35B/M02	3-speed man. w/floor shift (base six)	10.00		Esprit	5.00
35E/M20	4-speed man. (exc. Trans Am; w/34H, 34S only)	200.00	478/P05	Honeycomb wheels base, Formula	123.00
35G/M22	H-D 4-speed man. (Formula w/34S, 34X only)	231.00		Esprit	97.00
35J/M35	2-speed auto.		481/B80	Roof drop moldings (exc. Esprit)	15.00
	Base w/34D; n.a. in Calif.	174.00	484/B85	Belt reveal moldings (exc. Trans Am, req. w/582 w/534)	15.00
	Base, Esprit w/o 34D; n.a. Calif.	185.00	491/B96	Front & rear wheel opening moldings (base and Formula only)	21.00
35K/M38	Turbo Hydra-matic		492/B93	Door edge guards	6.00
	Base, Esprit, Formula w/o 34D	215.00	494/B84	Vinyl bodyside moldings (exc. Trans Am)	31.00
	Base w/34D	205.00	501/N41	Variable ratio power steering (exc. Trans Am; req. w/582, w/534)	113.00
35L/M40	Turbo Hydra-matic (Esprit, Formula w/o 34D, 34H)	236.00	502/JL2	Power front disc brakes (std. in Trans Am)	46.00
361/G80	Safe-T-Track differential (exc. Trans Am)	45.00	504/N33	Tilt steering (req. w/501; n.a. w/35J, K, L or w/o 431)	44.00
368/G92	Performance axle	10.00	521/B32	Front throw mats	7.00
401/U63	AM radio	65.00	522/B33	Rear throw mats	6.00
403/U69	AM/FM radio	135.00	524/B42	Trunk mat (incl. in 724)	8.00
405/U58	AM/FM stereo (incl. 411)	233.00	531/A01	All tinted glass	37.00
411/U80	Rear seat speaker	18.00	532/A02	Tinted windshield	30.00
412/U57	Tape player (req. w/401, 403, 405)	130.00	534/C49	Elec. rear window defogger (na w/34D, 541)	62.00
414/U55	Cassette player (na w/431, 424, 411, 412)	130.00	541/C50	Rear window defogger (na w/534)	31.00
423/K45	H-D air cleaner (exc. Trans Am)	9.00	551/A31	Power windows (req. W/431)	113.00
424/D58	Rear console (na w/414 w/431)	26.00	554/AU3	Power door locks	44.00
431/D55	Front console (na w/35A)	57.00	582/C60	Manual air conditioning	397.00
432/C24	Concealed wipers (base and Formula)	18.00	591/VJ9	California test (req. on Calif. cars)	15.00
434/D35	Body-color OSRV mirror (LH remote; n.a. w/444)	26.00	601/WU3	Hood air inlet (Formula req. 34X)	56.00
441/D34	RH visor vanity mirror	3.00	614/D98	Vinyl stripes (na w/494, Trans Am)	41.00
442/D31	Non-glare rearview mirror (incl. w/734)	6.00	611/D80	Rear air spoiler (Formula only)	32.00
444/D33	LH OSRV mirror (remote-control; n.a. w/434)	12.00	634/K65	Unitized ignition (req. w/34X on Formula)	77.00
451/AK1	Custom shoulder belts, front and rear	15.00	—/Y92	Convenience lamp group	11.00
461/N30	Custom cushion steering wheel (base)	15.00			

Sales Code and UPC	Description	Retail Price ($)
681/U05	Dual horns (base)	4.00
684/N65	Space-Saver spare tire (w/o 474, 478)	13.00
692/UA1	H-D battery (na w/582)	10.00
701/V01	H-D radiator	21.00
714/U30	Rally gauge w/tach & clock (na w/34D, 722, 718)	92.00
718/W63	Rally gauge cluster w/clock (na w/34D, 722)	46.00
722/U35	Electric clock	15.00
724/Y90	Custom trip group (Formula, Trans Am)	77.00

Sales Code and UPC	Description	Retail Price ($)
731/V30	Front and rear bumper guards	25.00
TGR/PX6	F78-14 whitewall fiberglass tires (base, Esprit)	44.00
THR/PL3	E78-14 whitewall fiberglass tires (base, Esprit)	28.00
TMF/PY6	F70-14 black fiberglass tires (base, Esprit)	35.00
TML/PL4	F70-14 white-letter fiberglass tires Base, Esprit	75.00
	Formula	40.00

Chapter 7

1973

★	Firebird Six
★	Firebird V-8
★	Esprit
★★	Esprit 400
★★	Formula 350
★★	Formula 400
★★★	Formula 455
★★★★	Formula SD-455
★★★★	Trans Am 455
★★★★★	Trans Am SD-455

Firebirds for 1973 had new colors, redesigned interiors, new hubcaps, and a longer option list. Bumpers were reengineered to meet federal standards, making the cars slightly longer. This resulted in a new grille that was slightly less recessed and had an egg crate pattern.

Among the year's technical changes were a new choke system and an exhaust gas recircula-

The 1973 Super-Duty heads are expensive to buy at swap meets. A variation on the Ram Air IV heads, they feature 1.77-inch exhaust valves and 2.11-inch intake valves with special cupped heads and swirl-polished finish. The free-flowing cast-iron round-port headers are designed for high performance, too. *F. Ann Billiamosa*

tion system. Hood scoops on Formulas and Trans Ams were sealed to keep water out and lower the noise level. Powerplants carried over compression ratios.

The base Firebird sold from $2,895 and looked the same as before in most regards. Front fenders carried Firebird lettering with a winged-bird emblem ahead of it. Smaller Pontiac lettering was set into the left-hand grille. When the V-8 was ordered, a red 350 call-out was added to the grille below the company name. The six had a new (8.2:1) compression ratio, 100 net horsepower at 3,600 rpm, and 175 foot-pounds of torque at 1,600 rpm. The 350-ci V-8s for the base Firebird were all two-barrel engines with 7.6:1 compression. They were rated for 150 net horsepower and 270 foot-pounds of torque and featured single exhaust.

There were several versions of the 350 and all other Pontiac V-8s—a total of 21 different engines in all. Some were certified for nationwide use, while others were approved for the sale either in California, in high-country regions, or in all states except California. Such things as transmission attachments, distributor model, and combustion chamber size varied between powerplants. Still other engine variations were the result of a midyear Environmental Protection Agency (EPA) order to remove a time-delay emissions system, which necessitated a virtual redesign of the system. All engines built after March 15 were modified, along with 700 earlier ones. Powerplants having the changes were painted a darker than normal blue. They also had two thermal valves tapped into the intake manifold and no exhaust gas recirculation system (EGR) solenoid.

Two changes in the basic features for 1973 were related to the redesigned interior. Lower door panels were molded instead of carpeted, and the woodgrain dash insert was switched from a flame chestnut pattern to African crossfire mahogany. Standard upholstery was the same as the previous year, but Custom trim was different. It featured deep contour bucket seats with three-bolster seatbacks set into a "horse collar" frame. Seat cushions were of a two-bolster design with the bolster inserts of perforated or Madrid-pattern Morrokide, Prado corduroy, or Bravo cloth.

Standard vinyl trims came in black, saddle, and white. Custom vinyl trims came in white, saddle, black, and burgundy. A rare option, available with white interiors, was nonstandard bright blue, orange, or red carpeting. The red carpeting was also available with black upholstery. There was also a beige-colored, cloth-and-vinyl combination.

For the exterior, color choices were expanded to 16, and all but one were brand new. Three—cameo white, Brewster green, and buccaneer red—were reserved for Trans Ams.

Esprit pricing began at $3,249. For distinction from the base Firebird, the Esprit had the same assortment of standard extras, plus dual horns and deluxe wheel covers (formerly a $5 option). Either all-vinyl or cloth-and-vinyl Custom trim was also standard.

The regular Esprit powerplant was the 150 net horsepower 350-ci V-8. Also available, at extra cost, was a two-barrel 400-ci engine with 8.0:1 compression, 170 net horsepower at 3,600 rpm, and 320 foot-pounds of torque at 2,000 rpm. Single exhausts were fitted and a Turbo Hydra-matic automatic transmission was required, at additional cost, on all Esprit 400s.

Formulas came in 350, 400, and 455 editions again, with the small-displacement version selling for $3,276. Late in the year, the powerful and rare Super-Duty (SD) 455 engine could be ordered for the Formula. It was the same as the Trans Am option, but the price was considerably stiffer at $675.

As in the past, a dual-scoop fiberglass hood characterized the Formula. The front fenders carried Formula lettering behind, an engine call-out, and a Firebird emblem. There was also a new, black-textured grille finish, high-rate rear springs, and the regular standard equipment that made the Formula a real performance machine. A new option was radial tires, also available for Trans Ams.

A plate denoting compliance to federal safety standards was attached to the driver's side doorjamb of all 1973 Firebirds. It had the vehicle identification number (VIN) at the bottom center, and the year and month of manufacture. This car was built in August 1973. *Mike Pergande*

Space-Saver spares were used on Firebirds from the start. A conventional spare was a no-cost substitute on early cars. However, to promote acceptance of the "temporary" spare, optional wheels and tires were priced lower for cars having the Space-Saver option. *Mike Pergande*

Dual exhausts were standard, even with the Formula 350, which was rated at 175 net horsepower at 4,400 rpm. This engine developed 280 foot-pounds of torque. Specs for the Formula 400 were 230 net horsepower at 4,400 rpm and

Trans Am sported a Formula steering wheel (not standard equipment), swirl-finish instrument panel trim plate, Rally gauge cluster with clock and tachometer, and Firebird emblem in center. This car is one of only 72 SD-455s built with four-speed transmission.

325 foot-pounds of torque. The non-SD 455-ci engine came several ways, depending on sales area certification and transmission choice. One version had treaded rocker arm studs. Different cylinder heads gave three different combustion chamber volumes. The 455 had 8.0:1 compression and was rated for net horsepower at 4,000 rpm and 370 foot-pounds of torque at 2,800 rpm.

Like others before it, the 1973 Trans Am had all of the special air dams, spoilers, flares and scoops, although the hood scoops were sealed and nonfunctional. Besides having an expanded list of exterior finish colors, the 1973 Trans Am was the first to offer an optional "big bird" hood decal that was much larger than the standard bird decal on the nose. The big bird was always black, but the background color varied: it was orange on red cars, black on white cars, and light green on the Brewster green Trans Ams.

The base engine was the non-SD 455 cubic inch V-8 also used in Formulas. The SD 455 cubic inch V-8 was also used in Formulas and was an on-again, off-again proposition that finally became an official Formula and Trans Am option in the spring. However, experts say it was not until July that the first SD cars reached the showroom.

The SD 455 engine evolved from PMD's Trans Am racing program of the early 1970s. It represented a low-compression, extra-horsepower

Trans Ams featured a rear decklid spoiler, model identification decal, and chrome exhaust-pipe extensions at the rear. Car shown here is a Buccaneer red SD-455 owned by *Bob Adams of Illinois.*

option made available in limited quantities. It featured a special block with reinforced webbing, large forged-steel connecting rods, special aluminum pistons, a heavy-duty oiling system (with dry sump pump provision), a high-light camshaft, and four-bolt main bearing caps, plus a special intake manifold, dual exhaust system and valvetrain components.

Numerous enthusiast magazines tested the SD-455 and came up with outstanding performance figures. In a *Hot Rod* report, a SD-455 Trans Am turned in a 13.54 second quarter-mile at 104.29 miles per hour. *Car and Driver* results were quite similar with a 13.75-second, 103.56-mile-per-hour performance. This engine has also become a legend with Trans Am lovers, and 1973 models with the black SD-455 lettering on the hood scoop will bring up to twice as much money as a regular 455 Trans Am!

In his "collecting" column in the June 1986 issue of *High-Performance Pontiac* magazine, Jerry Heasley reported on the sale of a white 1973 Trans Am SD-455 for $25,000. The car was

Standard Firebird interior, shown here in black, was standard on base coupes, Formulas, and Trans Ams, but not available for Esprits. The rear console was a separate option that's considered rare for early Firebirds. Note the pattern of seat inserts and inner door panels of the standard interior. *Bob Adams*

Factory photo shows 1973 SD-455 with white-stripe tires. Standard equipment called for F60-15 white-letter tires on 15-inch Rally II wheels. The GR70-15 steel-belted radial whitewalls were an extra. Honeycomb wheels could be substituted for these standard wheels at no extra cost. *Pontiac Motor Division*

owned by Mark Turpin, who purchased it new, and put only 15.8 miles on the odometer. Turpin sold it after a Christie's auction in New York city in early 1986. It had the four-speed manual transmission, 310 horsepower engine, Custom interior, tilt steering wheel, and Honeycomb wheels. The figure of $25,000 appears to be the record for an SD-455 at this time.

Pontiac made no basic changes in its method of coding the 1973 Firebird VINs. The fifth character again indicated which powerplant was correct for a car, while the sixth was changed to a 3 for 1973. The engine identification codes were as follows:

Engine	Displacement	Carburetor	Horsepower	Exhaust	Code	Compression
L-6	250 ci	1V	100	single	D	8.2:1
V-8	350 ci	2V	150	single	M	7.6:1
		2V	175	dual	N	
V-8	400 ci	2V	170	single	R	8.0:1
		2V	170	dual	P	
V-8	400 ci	4V	230	single	S	8.0:1
		4V	230	dual	T	
V-8	455 ci	4V	250	dual	Y	8.0:1
		4V	310	dual	X	8.4:1

Two-character engine codes stamped on the blocks indicated the type of engine in the car and the correct transmission attachment. For V-8s, the codes were:

Displacement	Carburetor	Transmission	Code
350 ci V-8	2V	manual	ZB
		automatic	WA, WL, ZW
400 ci V-8	2V	manual	—
		automatic	X4, X3, W5
	4V	manual	Y6, YG
		automatic	XN, XX, YK
455 ci V-8	4V	manual	ZZ, ZE
		automatic	XE, XL, X7, XM
SD-455 ci V-8	4V	manual	ZJ
		automatic	XD

The two-character block codes also revealed compression ratio, combustion chamber volume, valve type, camshaft, lifter type, rocker arm stud design, distributor model, usage (certification), and whether the engine was an early-year version with the spark delay system.

Firebird production began a steady, upward climb in 1973, bearing out GM's wisdom in keeping the car on the market. By the end of the upward trend, Firebird sales accounted for over 175,000 units per year and over 23 percent of PMD's total business. It also became a very profitable car line, with 55 percent of all Firebirds being delivered as Trans Ams by 1979. For 1973, the Firebird production totals were:

Name	Code	Manual	Automatic	Total
Base	FS87	—	—	14,096
Esprit	FT87	—	—	17,249
Formula	FU87	—	—	10,123
Formula SD-455	FU87	—	—	43
Trans Am 455	FV87	1,420	3,130	4,550
Trans Am 455-SD	FV87	72	180	252
Total				46,313

In its 1975 Buying Guide, *Consumer Reports* gave the 1973 Firebirds the highest reliability rating that it had ever recorded for the marque. Based on owner reports covering the frequency of repairs needed in 17 areas, the publication showed that the 1973 models required only the average number of repairs to all systems and in all body service categories. In addition, the Firebirds were found to require fewer-than-average repairs overall. The only other cars rated as highly were the 1973 Oldsmobile Tornado (among domestic models), and the Fiat Spider and Porsche 914 (among imports).

A second publication, *Consumer Guide's Used Car Rating and Price Guide 1971–1980* lumped the 1973 Firebirds into a category with

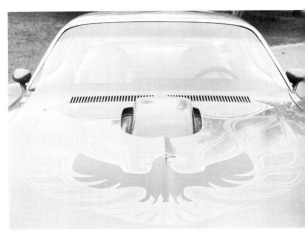

The standard Trans Am decal was a small, but colorful, "chicken" adorning only the nose of the car. It differed from the decal used on the 1970 1/2 through 1972 models. *Bob Adams*

GM stylist John Schinella overcame corporate resistance to have his large "chicken" decal accepted as a Trans Am production option. This particular design was used through 1978. *Pontiac Motor Division*

The 1973 Firebird sales catalog pictured an Esprit coupe with Navajo orange finish on its cover. This model continued to be externally distinguished by dual body-color sport mirrors, body-color door handle inserts, extra bright body moldings and a model identification script on the roof sail panels. The deluxe wheel covers shown here were also standard equipment. Grille and fender badges indicate that the base (for Esprit and Formula) 350-ci V-8 is in this car.

all pre-1976 models. The overall group was said to have a slightly worse than average repair rate.

According to this source, there were continuing complaints about early rusting, poor paint finish, and loose bodies. However, the 1973 models were not discussed separately.

Firebird collectors tend to favor the higher rating accorded by *Consumer Reports*. They insist that the 1973 models are virtually trouble free and require no more than the normal amount of service. There were no factory recalls on 1973 Firebirds, although recalls were made on both 1972 and 1974 models.

As far as performance characteristics, there was an obvious cutback in performance in cars not equipped with the Super-Duty engine option. For example, *Road & Track* editors tested a 1973 Esprit with the 400-ci V-8 that produced 170 net horsepower. The test car was equipped with Turbo Hydra-matic transmission and a 2.73:1 rear axle. It produced 0–60 time of 11.2 seconds and got through the quarter-mile in 19.71 seconds with an 86-mile-per-hour terminal speed.

These figures were somewhat comparable to the performance of earlier ohc-six-powered Firebirds and earlier Esprits with the smaller 350-ci V-8 engine. They were also quite a bit under the figures for previous 400-ci V-8-powered Firebirds.

From a collector's point of view, the six low-rung 1973 models are not particularly attractive investments. These cars should appreciate in value at slightly under average rates. However, because of that improved reliability record, they make fine cars for everyday driving use.

The Formula 455 model is slightly more desirable to collectors, although its year-to-year appreciation potential seems to be only about average. You can buy and use this type of car and still come out about even.

More desirable to collectors would be a Trans Am with the standard 455 engine, or one of the 43 Formulas sold with the SD-455 option. Both of these models should appreciate at higher-than-normal rates and represent good investments.

The top investment car of the year is the SD-455 Trans Am. Combining both the special Trans Am appearance package and the high-performance engine, this car was also made in extremely limited numbers. Asking prices for SD-455 T/As range between $12,000 and $20,000 today and appear to be moving rapidly upward from there. Four-speed manual

The base Firebird coupe had narrower rocker panel moldings and small, smooth hubcaps that looked like the "baby moons" popular with hot rodders in the fifties. Use of bright body edge moldings was limited to the windshield, rear window, and grille, unless optional trim was added. The dual sport mirrors and 350 grille and fender call-outs indicate that the Admiralty blue coupe shown in sales literature had a couple of additional options.

The sales catalog for 1973 showed the standard interior, in black, of a car with the Formula steering wheel and automatic-transmission type of gear shifter and console.

The Formula was easily identified by its special, twin-snout fiberglass hood, a black-textured grille, dual sport mirrors, and engine call-outs. This car has extra-cost Rally II wheels and white-letter Firestone wide ovals.

transmission attachments are particularly rare and have much more appeal to collectors.

There doesn't seem to by any real "sleeper" models for 1973, as the general rise in production of all standard models makes them more common than earlier Firebirds of the same car line. There doesn't seem to be any particularly rare option combinations for the non-Super-Duty models, either.

1973 Firebird Options and Accessories

Sales Code and UPC	Description	Retail Price ($)
542/K45	H-D air cleaner (na w/Trans Am, SD-455)	9.00
582/C60	Custom air conditioning (na w/base six)	397.00
691/TP1	Maintenance-free battery	26.00
692/UA1	H-D battery (na w/air conditioning)	10.00
502/JL2	Power disc brakes (std. w/Trans Am)	46.00
614/24F	Bright blue carpet (w/trims 321, 421)	n/c
612/97F	Orange carpet (w/trims 321, 4210)	n/c
611/75F	Red carpet (w/trims 521, 361, 421, 461)	n/c
711/U35	Electric clock (w/rally gauges; std. w/Trans Am)	15.00
431/D55	Front console (w/floor shift only)	57.00

Sales Code and UPC	Description	Retail Price ($)
424/D58	Rear console	26.00
SVT/C08	Cordova top	
	Base Firebird and Formula	87.00
	W/Esprit; n.a. w/Trans Am	72.00
512/WW7	Trans Am hood decal (Trans Am only)	55.00
541/C50	Rear defogger (na w/elec. defrost.)	31.00
534/C49	Elec. Rear defogger (na w/6 or elec defrost.)	62.00
371/G80	Safe-T-Track rear axle (std. w/Trans Am)	45.00
554/AV6	Elec. door locks	44.00
712/W63	Rally gauges w/clock (na w/6, Trans Am)	46.00

Sales Code and UPC	Description	Retail Price ($)
714/WW8	Rally gauges w/tach. (na w/6, 712; std. w/Trans Am)	92.00
531/A09	All windows tinted	37.00
532/A02	Tinted windshield	30.00
492/B93	Door edge guards	6.00
732/V32	Rear bumper guards (incl. w/342)	15.00
701/K05	Engine block heater	10.00
634/WU3	Ram Air hood (w/Formula 400, Formula 455)	56.00
681/U05	Dual horns (base Firebird only; std. w/other models)	4.00
694/K65	Unitized ignition w/400-4V. 455-4V in Formula and Trans Am)	77.00
534/AU3	Elec. door locks	44.00
604/B42	Trunk mat (std. w/Esprit & Custom trims)	8.00
621/B32	Front floor mats	7.00
622/B33	Rear floor mats	6.00
441/D34	RH visor vanity mirror	3.00
434/D35	Dual body-color OSRV mirror (LH remote; base Firebird)	26.00
491/B96	Front & rear wheelhouse moldings (std. w/Esprit; n.a. w/Trans Am)	15.00
481/B80	Roof drip moldings (na w/Trans Am)	31.00
484/B85	Window & rear hood moldings (std. w/Esprit)	21.00
701/U01	H-D radiator (na w/air conditioning)	21.00
411/U63	AM radio	65.00
413/U69	AM/FM radio	135.00
415/U58	AM/FM stereo (na w/421; req. w/front console)	233.00
451/AK1	Custom seat belts (incl. w/front shoulder straps)	15.00
684/N65	Space-Saver spare tire (n/c w/478, 474, VJ)	15.00
421/U80	Rear speaker (na w/415)	18.00
632/D80	Rear deck lid spoiler (std. w/Trans Am)	113.00
461/N30	Custom cushion steering wheel (base Firebird only)	15.00
464/NK3	Formula steering wheel w/501 only in base Firebird	56.00
	w/501 only in Formula, Esprit)	41.00
504/N33	Tilt steering wheel (na w/column shift)	44.00
631/D98	Accent stripes (na w/Trans Am)	41.00
422/U57	Stereo tape player (req. w/radio & 431)	130.00
421/P06	Wheel trim rings (incl. w/Rally II rims)	26.00
724/Y90	Custom trip group*	77.00
422/P02	Custom finned wheel covers on Esprit na w/342	24.00
	base Firebird, Formula; n.a. w/342	50.00
476/P01	Deluxe wheel covers (base Firebird, Formula w/342)	26.00
478/P05	Honeycomb wheels base Firebird/Formula w/o 684, GR70-15 tires	123.00
	Formula w/684 or GR70-15 tires	103.60
	Esprit	97.00
	Trans Am	n/c
474/N98	Rally II Wheels base, Esprit, Formula w/o 684 GR70-15 tires	87.00
	Formula w/684, GR70-15 tires	74.80
	Esprit	61.00
551/A31	Power windows (w/front console only)	75.00
432/C24	Concealed windshield wipers (base Firebird & Formula)	18.00

A redesigned Custom interior option was new for 1973. It featured "breathable" pigskin vinyl inserts for seats and door panels. Also shown here is the Custom cushion steering wheel, automatic transmission (and console), and console-mounted power window controls. The Custom interior and steering wheel were standard in Esprits; optional in other Firebirds.

Sales Code and UPC	Description	Retail Price ($)
331/Y88	Basic option group**	
	base Firebird 6-cyl.	441.00
	base Firebird w/350 V-8	451.00
	Esprit	421.00
	Formula 350	459.00
	Formula 400/455	480.00
	Trans Am	65.00
334/Y95	Body protection group (exc. Trans Am†)	52.00
344/Y92	Lamp group††	10.00
342/Y99	Formula handling package, Formula only	
	w/F60-15 tires & 684	72.60
	w/F60-15 tires & w/o 684	87.00
	w/G70-15 tires & w/o 684	157.00
	w/G70-15 tires & 684	150.60
35D/Std.	6-cyl. engine (na w/air conditioning)	n/c
35M/L30	350 V-8 2-bbl (req. w/Turbo Hydra-matic in Calif.)	118.00
35R/L65	400 V-8 2-bbl (req. w/Turbo Hydra-matic at extra cost)	51.00
35S/L78	400 V-8 4-bbl. (req. w/Turbo Hydra-matic or 4-speed; Formula)	97.00
35W/L75	455 V-8 4-bbl (Turbo Hydra-matic or 4-speed only; std. w/Trans Am)	154.00
35X/LS2	SD-455 V-8 4-bbl (req. W/Turbo Hydra-matic or 4-speed and Hood air inlet	
	Formula	675.00
	Trans Am	521.00
36L/M38	Turbo Hydra-matic	
	base Firebird six	205.00
	base 350, Esprit 350, Formula 350	215.00
36B/M12	3-speed man. trans. w/floor shift (std. w/base six, Esprit 350)	n/c
36E/M20	4-speed manual (std. w/Trans Am; n.a. w/350, 400-2V)	200.00
36H/M21	4-speed manual; close-ratio (Formula 2/400, 455-4V)	200.00
371/G80	Safe-T-Track axle (std. w/Trans Am)	45.00
378/G92	Performance axle (all)	10.00
	E78-14 whitewall fiberglass tires (na w/342 on 2Y or 2Z)	
HR/PL3	all base Firebird & Esprit w/o 684	28.00
HR/PL3	all base Firebird & Esprit w/684	22.40
	F78-14 blackwall fiberglass tires	
GF/PX5	all base Firebird & Esprit w/o 582, 2/684	14.00
GF/PX5	all base Firebird & Esprit w/o 582, w/684	11.20

Sales Code and UPC	Description	Retail Price ($)
	F78-14 whitewall fiberglass tires	
GR/PX6	all base Firebird & Esprit w/582, 684	24.00
GR/PX6	all base Firebird & Esprit w/582, w/o 684	30.00
GR/PX6	all base Firebird & Esprit w/o 582, w/684	44.00
	F70-14 blackwall fiberglass tires	
MF/PY6	all base Firebird & Esprit w/582, 684	16.80
MF/PY6	all base Firebird & Esprit w/582, w/o 684	21.00
MFA/PY6	all base Firebird & Esprit w/o 582, w/o 684	28.00
MF/PY6	all base Firebird & Esprit w/o 582, 684	35.00
	F70-14 white-letter fiberglass tires	
ML/PL4	all Formula w/o 342 & w/684	32.00
ML/PL4	all Formula w/o 342 & 684	40.00
ML/PL4	all base Firebird & Esprit w/582, 684	48.80
ML/PL4	all base Firebird & Esprit w/582, w/o 684	51.00
ML/PL4	all base Firebird & Esprit w/o 582, w/684	60.00
ML/PL4	all base Firebird & Esprit w/o 582, w/684	75.00
	GR70-14 whitewall steel-belted radial	
VJ/P85	Trans Am w/o 684	70.00
VJ/P85	Trans Am w/684	56.00

*Custom trim group includes deluxe front bucket seats, assist straps on door and glovebox, roof insulator pads, formed vinyl trunk mat, body-color outside door handles, and rear quarter ashtray (standard in Esprit, available in Formula and Trans Am).

**Basic option group includes Turbo Hydra-matic, whitewall tires (white-letter tires, Formula), AM radio, wheel trim rings (standard, Trans Am; deluxe wheel covers, Esprit), dual horns (base Firebird), and power steering (standard, Trans Am).

†Body protection group includes door edge guard, vinyl bodyside moldings and rear bumper guards.

††Lamp protection group includes luggage lamp, glovebox lamp, and interior panel courtesy lamp.

‡Formula handling package (Formula only) includes F60-15 white-letter tires or G70-15 whitewall radial tires, heavy-duty rear springs, heavy-duty shock absorbers, heavy-duty front and rear stabilizer bars, and 15x7-inch wheel rims.

Chapter 8

1974

★	**Firebird Six**
★	**Firebird V-8**
★	**Esprit**
★★	**Esprit 400**
★★	**Formula 350**
★★	**Formula 400**
★★★	**Formula 455**
★★★★	**Formula SD-455**
★★★★	**Trans Am 455**
★★★★★	**Trans Am SD-455**
★★★★	**Trans Am 400**

New styling and better sales were part of the 1974 Firebird story. There were changes in the engines used in these cars and the SD-455 option remained in limited availability. Four basic models were offered again and each could be equipped with a myriad of options.

Designer John Schinella developed a more integrated "soft" bumper treatment for the front and rear of the 1974 models. It featured a construction of urethane foam bonded to steel frame horns for improved impact protection.

The front of the cars had a slanted, "shovel-nose" grille cap incorporating an "electric-shaver" grille with slanting vertical blades. Black rubber-bumper-face bars were featured. An air-scoop-like front lower valence gave the front of the car a more massive look. The parking lamps were slimmer and wider, with the chrome protective grids of earlier years eliminated. They carried textured, amber-colored lenses.

A soft, urethane end cap was also used at the rear of 1974 models. It incorporated wider, horizontal-louvered taillights. The rear fender line was completely lowered to go with the new, wrap around-style soft trim. Pontiac said that the new rear styling was engineered "for added strength

The 1974 Firebird was a transitional model with a new front treatment, but its type of rear window was used since 1970 1/2. It listed a few less horse-power on paper, however. Patterning of the standard seats, also shown, was unchanged. *Pontiac Motor Division*

against minor impacts."

A dramatic increase in Firebird sales (up 27,416 units) was particularly strong for the performance-oriented Trans Am, which saw a 113-percent gain in deliveries. This was brought about primarily by the disappearance of competitors like the then-comparable Chevy Camaro Z/28, Ford Boss Mustang, Plymouth Barracuda, Dodge Challenger R/T, and AMC AMX. In fact, the Trans Am and the Corvette were the sole survivors in the domestic performance car market, and the hot Pontiac sold for about half as much as Chevrolet's sports car.

Firebird interiors for 1974 were carried over from the previous year, except for changes in color schemes. Buyers could order multicolor interior combinations such as white seats with red door panels, dashboard, and carpeting. Also, the fabric formerly called Bravo bolster cloth was renamed Bravado bolster cloth.

There were again, 16 exterior color combinations including the following new colors: Chestwood brown, Fernmist green, Fire Coral bronze, Pinemist green, Gulfmist aqua, Limefire green, Carmel beige, Colonial yellow, Sunstorm yellow, Honduras maroon, and Denver gold. Cameo white, Buccaneer red, and Admiralty blue were offered only for Trans Ams. (Admiralty blue replaced Brewster green, which was dropped.)

The base Firebird retailed for $3,174.70. Its trim consisted of Pontiac lettering on the left-hand grille, Firebird front fender side lettering, and a bird emblem on the rear deck. Cars with the optional V-8 had red 350 callouts under the fender lettering.

A reduced 8.2:1 compression ratio gave the 1974 six a new output rating of 100 net horsepower at 3,600 rpm. The torque figure went to 175 foot-pounds at 1,600 rpm. An optional 350-ci V-8, with two-barrel carburetion and single exhaust, continued to come with a 7.6:1 compression ratio. Its horsepower and torque ratings were slightly increased to 155 net horsepower at 3,600 rpm and 275 foot-pounds at 2,400 rpm.

Remaining as the Firebird series "luxury" model was the $3,526.70 Esprit. It again came with Esprit sail panel scripts, body-color sport mirrors, concealed wipers, color-keyed door handles, extra trim moldings, deluxe wheel covers, and Custom interior trim.

The single-exhaust 350-ci V-8 was the standard engine in the Esprit. Buyers could also pay $51 extra for a 400-ci V-8 with a two-barrel carburetor. It produced 175 net horsepower at 3,600 rpm and 315 foot-pounds of torque at 2,000 rpm.

Selling for $3,614.20, the Formula package for 1974 included the normal street-performance goodies: dual-scoop fiberglass hood, Custom cushion steering wheel, black textured grille, dual

You can see the new, slanted shovel-nose soft front end and rear body-color panel with rubber bumper even in this profile view of the 1974 SD-455 Trans Am. Regular Trans Ams came with either 400- or 455-ci, non-Super-Duty engines. All Trans Ams were available with the small bird decal on nose or larger bird decal on hood.

exhaust, and so on. In its least potent form, with a 350-ci V-8, the Formula provided 170 net horsepower at 4,000 rpm and 290 foot-pounds of torque at 2,400 rpm.

A two-barrel, 400-ci V-8 was also a $51 option for Formulas. This version of the engine featured dual exhausts and more power. It developed 190 net horsepower at 4,000 rpm and produced 330 foot-pounds of torque at 2,400 rpm. Or, for $97, the buyer could order a four-barrel formula 400 (L78 option) engine that produced 225 net horsepower at 4,000 rpm and 330 foot-pounds of torque at 2,400 rpm.

There was also a pair of 455-ci engines available for 1974 Formulas. The L75 version was a four-barrel, dual-exhaust option with 250 net horsepower at 4,000 rpm and 380 foot-pounds of torque at 2,800 rpm. The top—and rarest—engine was the LS2 option: the SD-455 with a new, 8.4:1 compression ratio. It produced 290 net horsepower at 4,000 rpm and 395 foot-pounds of torque at 3,200 rpm. Only 58 Formulas were built with this engine during the 1974 model run. They are much rarer than SD-455 Trans Ams, although not necessarily more valuable, because the Trans Ams came with other extras.

The 1974 Trans Am package was about the same as in previous model years. It included all standard Firebird equipment plus a Formula steering wheel, Rally gauges with a clock and dash panel tachometer, swirl-finish dash panel

trim plate, full-width rear deck lid spoiler, power steering and front disc brakes, limited-slip differential, wheel-opening air deflectors, front fender air extractors, dual exhaust with chrome extensions, Rally II wheels with trim rings, special suspension, dual outside rear view sports mirrors, F60-15 white-letter tires, a four-speed manual transmission, and the 400-ci, 225 net horsepower V-8. Both the standard and SD-455 cubic inch engines were available at extra cost above the base price of $4,350.75

Pontiac again made no basic changes in its method of coding Firebirds. The fifth character in the VIN indicated the type of engine used in a car. The sixth character was changed to a 4 for 1974. The engine identification codes were as follows:

Engine	Displacement	Carburetor	Exhaust	Code
L-6	250 ci	1V	single	D
V-8	350 ci	2V	single	M
			dual	N
	400 ci	2V	single	R
			dual	P
		4V	single	S
			dual	T
	455 ci	4V	dual	Y
	SD-455 ci	4V	dual	X

Two-character engine codes stamped on the blocks indicated the type of engine in the car and the correct transmission attachment. For 1974, the codes were:

On the 1974 Formula, Rally II rims and wide-letter tires were extra-cost equipment. Factory-style tires of white-letter design now read Firestone 500 instead of Firestone Wide-Oval. Both Formulas and Trans Ams were promoted as "Pontiac's sports cars."

Engine	Displacement	Compression	Net SAE Horsepower	Transmission	Code
L6	250 ci	8.2:1	100	manual	AA
				automatic	CCR
V-8	350 ci	7.6:1	155/170	manual	WA, WB
				automatic	AA, YA, YB, YC, YN, YP, YS, ZA, ZB
	400 ci	8.0:1	175/190*	manual	WN, WP
				automatic	AH, YH, YJ, ZH, ZJ
			175/190*	manual	—
				automatic	AD, YF, YK, ZD, ZK
			225*	manual	WT
				automatic	AT, YT, YZ, ZT
			225**	manual	WR, Y3
				automatic	AD, YL, YM, ZS
	455 ci	8.0:1	215/250	manual	—
				automatic	AU, YR, YU, YAW, YX, YY, ZU, ZW, ZX
	455 ci (SD)	8.4:1	290	manual	W8
				automatic	Y8
	455 ci	8.0:1	215/250**	manual	—
				automatic	AH, Y4, Y6, Y9, Z4, Z6

*Early-production engines
**Late-production engines

The SD-455 engine option was used in relatively few Formulas and Trans Ams, but it was widely publicized that cars with this powerplant offered higher performance than any other domestic 1974 model, including the more expensive Corvette. This did a great deal for the Firebird's high-performance image and brought many new buyers into Pontiac showrooms. When coupled with the discontinuance of other high-

The Custom trim option shown here was standard in Esprits but cost extra in other models. This car has the three-spoke Custom cushion steering wheel and "automatic" type of gear shifter and console. It came in saddle, red, white, blue, green, and black combinations.

All models, except Trans Ams, had a woodgrain instrument panel insert. The Formula steering wheel depicted here was 14 inches in diameter and optional in all Firebirds, except Trans Ams. This car also sports the Rally gauge cluster, another standard option in Trans Ams but extra in other Firebirds.

performance pony cars, this image enhancement had a very positive effect on Firebird sales.

Available production totals for the 1974 models and equipment packages were as follows:

Name	Engine	Code	Manual	Automatic	Total
Base Firebird	6-cyl	FS87	—	—	7,603
	V-8	FS87	—	—	18,769
Esprit		FT87	—	—	22,583
Formula Firebird	all exc. SD-455	FU87	—	—	14,461
	SD-455	FU87	—	—	58
Trans Am	400 ci V-8	FV87	1,750	2,914	4,664
	455 ci V-8	FV87	—	4,648	4,648
	SD-455 V-8	FV87	212	731	<u>943</u>
Total					73,729*

*64,403 Firebirds were sold with automatic transmission; 4,543 Firebirds with four-speed manual transmission; and 4,783 Firebirds with three-speed manual transmission.

The repair rate for 1974 Firebirds dropped back to the slightly worse than average category. This was partly because of a problem with the emissions system that prompted a factory recall for cars with 350- or 400-ci V-8s with two-barrel carburetion. The problem was the possibility of a defective thermal vacuum valve, which activated the exhaust gas recirculation (EGR) system.

Despite this minor trouble, the Firebirds were highly rated by the editors of several car magazines. *Road Test* called the Firebird a "dynamite" car and described the Trans Am as "super." *Motor Trend* nominated the Firebird for its Car of the Year award and called it, "The best combination of pure performance and handling in the U.S. Grand Touring market . . . the sole survivor of a vanished era."

In one survey, *Motor Trend* asked 1,500 Japanese car enthusiasts to pick the best car in the world. The Firebird came out the overall favorite choice, beating the Mustang by a slight margin and the Corvette by a wider one.

Several technical changes in 1974 models have directly affected their appeal to car collectors today. They pertained to fuel economy, tires, and the availability of the two 455-ci high-performance engine options.

Two midyear changes were designed to enhance fuel economy characteristics after the Arab oil embargo focused attention on higher gasoline prices and made gas temporarily hard to get. The first was the introduction of a new, high-

This rear view of the base Firebird coupe for 1974 illustrates the year's wider taillights and redesigned rear bumper system. Options shown here are white-stripe tires, Rally II wheels, and dual sport mirrors. The front fender nameplates also have 350 V-8 engine call-outs below them. In addition, the door handles have body-color inserts, which means the Custom interior option has been ordered too.

The Esprit had wider rocker panel moldings than the base Firebird coupe. The deluxe wheel covers were standard, too. The Esprit name appeared, in bright metal signatures, on both rear roof quarters (sail panels), and the door handles had color inserts to go along with the standard-equipment Custom interior. A V-8 was again standard in this model: the 350-ci, two-barrel V-8.

energy ignition (HEI) system for Firebirds with the 400-ci V-8 and the non-SD 455-ci V-8. The second was the release of a new fuel economy gauge package as an extra-cost option.

In the area of tires, a GM corporate edict specified that all 1974 GM cars, including Firebirds, would be sold with steel-belted radial tires. The new radial tuned suspension (RTS) system gave overall better ride quality and improved the Firebird's adhesion to wet road surfaces. However, the new steel-belted radials did not have the same degree of dry-surface cornering ability as the older, nine-inch-wide F60-15 bias-belted tires.

For enthusiasts, a more negative mid-1974 model year change was the discontinuance of both the standard 455-ci V-8 and the SD-455 engine options. According to Mike Lamm in his book, *The Fabulous Firebird,* "Both 45 5 V-8s got quietly dropped from the Firebird option lists during 1974, partly because big engines weren't selling at all after the oil crunch and also because dropping them simply seemed good pubic relations."

In today's collector market, the desirable 1974 Firebirds would probably be the Formula SD-455 and the Trans Am SD-455, especially examples with a four-speed manual transmission.

However, the Trans Am with the 400-ci V-8 and manual transmission could be a "sleeper" car because of its relatively low production total of just 1,750 units.

Finned wheel covers (C) were available at extra-cost on all Firebirds except Trans Ams. Rally II wheels (D) were standard on Trans Ams and available on all other models, while Honeycomb wheels (E) were available on all Firebirds.

As revealed by the engine call-outs on the scoop, this car is a 1974 Formula with the optional SD-455 engine, but the twin-scoop hood was deleted in favor of the Trans Am setup with single, rear-facing shaker scoop. Gary D. Menen is the owner of this rare Pontiac. *Automobile Quarterly*

1974 Firebird Options and Accessories

Sales Code and UPC	Description	Retail Price ($)
604	H-D dual-stage air cleaner (na w/V87, 35X)	9.00
582	Custom air conditioning (na w/35D)	446.00
378	Performance axle (req. w/682 w/manual trans.)	10.00
371	Safe-T-Track axle (std. w/V87)	45.00
574	Maintenance-free battery (na w/591)	26.00
591	H-D battery (na w/574)	10.00
502	Power front disc brakes (req. w/35X)	46.00
711	Electric clock (incl. w/V87, 712, 714)	15.00
431	Front console (na w/36A)	58.00
424	Rear console	26.00
594	Rear window defogger (na w/592)	33.00
592	Elec. rear window defroster (na w/35D, 594)	64.00
682	Dual exhaust (na w/35D; std. w/U87, V87)	45.00
571	Soft-Ray glass, all windows	38.00
572	Soft-Ray glass, windshield only	31.00
492	Door edge guards	6.00
601	Engine block heater (na w/35D)	10.00
512	V87 hood decal (V87 only)	55.00
694	Dual horns (S87 only, std. w/other Firebirds)	4.00
602	Unitized ignition (w/35S, W only; n.a. w/S87, T87)	77.00
654	Lamp package (luggage, glovebox, Instrument panel courtesy)	10.00
554	Elec. door locks	46.00
622	Front floor mats	7.00
624	Rear floor mats	6.00
614	Trunk mat (std. w/Custom buckets and T87)	8.00
434	Dual sport mirrors (std. w/T87, U87, V87)	26.00
494	Black vinyl bodyside moldings (na w/V87)	31.00
481	Roof drip moldings (std. w/T87)	15.00
491	Wheel opening moldings (std. w/T87; n.a. w/V87)	15.00
484	Window sill & rear hood edge Moldings (std. w/T87)	21.00
511	Custom pedal trim plates (std. W/T87)	5.00
411	AM push-button radio	65.00
413	AM/FM radio	135.00
415	AM/FM stereo	233.00
712	Rally gauges and clock (na w/714, 711, V87)	46.00
714	Rally gauges w/clock and tach. (na w/711, 712; std. w/V87)	92.00
514	Ram Air hood (U87 w/35S only; incl. w/35X; std. w/V87)	56.00
691	Custom front & rear seat/shoulder belts	15.00
421	Rear speaker (na w/415, 422; 411 or 413 req.)	21.00
681	Rear air spoiler (std. w/V87; n.a. w/T87)	na
461	Custom cushion steering wheel (std. w/T87, U87; n.a. w/V87)	15.00
464	Formula steering wheel Std. w/V87; 501 req.	41.00
	S87; 501 req.	56.00
504	Tilt steering wheel (na w/column-shift Hydra-matic)	45.00
638	Vinyl tape stripes (na w/V87)	41.00
422	Stereo 8-track player (431 w/411, 413, 415 req.)	130.00
684	Space-Saver spare tire	na
472	Custom finned wheel cover T87	24.00
	na w/V87	50.00
476	Deluxe wheel covers (std. w/T87; n.a. w/V87)	26.00
471	Wheel trim rings (std. w/V87; n.a. w/T87)	26.00
474	Rally II wheel rims S87 & U87 w/o 684	87.00
	U87 w/684	74.80
	T98	61.00
478	Honeycomb wheel rims S87 & U87 w/o 684	123.00
	U87 w/684	103.00
	T87	97.00
551	Power windows (w/431 only)	78.00
432	Concealed wipers (std. w/T87, V87)	18.00
541	Calif. emissions test (req. w/Calif. cars)	20.00
SVT	Vinyl top T87	72.00
	S87, U87, V87	87.00
734	Impact protection bumper	5.00
722	H-D radiator (na w/582)	21.00
804	Kilo speedometer	20.00
35D	250 ci 6-cyl. (S87)	std.
35M	350 ci 2V (std. w/T87, U87; n.a. w/V87)	118.00

Sales Code and UPC	Description	Retail Price ($)
35R	400 ci 2V (na w/S87, V87; w/36L only)	51.00
35S	400 ci 4V (w/36L, 36E only; n.a. w/S87, T87)	97.00
35W	455 ci 4V	
	V87 price only	57.00
	U87 price only; 36L req.	154.00
35X	SD-455	
	V98 price only; req. w/36E, 36L	578.00
	U87 price only; req. w/36E, 36L)	675.00
36B	3-speed man. trans (w/floor shift; w/35D, 35M only)	n/c
36L	M-40 Turbo Hydra-matic (na w/35D, 35R, 35W)	242.00
36E	4-speed man. trans. (na w/35D, 35R, 35W)	207.00
36K	M-38 Turbo Hydra-matic	
	w/base S87 6-cyl.	221.00
	w/T87, U87	211.00
308	Custom trim package	77.00
TGJ*	Radial tuned suspension	
	T87 w/o 582, 684	145.00
	T87 w/582, 684	119.00
	T87 w/582, w/o 684	107.80
	U87 w/o 684	30.00
	U87 w/684	25.00
TGK**	Radial tuned suspension	
	S87 or T87 w/o 582, 684	155.00
	S87 or T87 w/582, 684	127.00
	S87 or T87 w/582, w/o 684	141.00
	S87 or T87 w/582, w/684	115.80
TGR	F78-14 whitewall fiberglass tires (Firebird, Esprit)	
	w/o 582, 684	44.00
	w/582, 684	35.20
	w/582, w/o 684	30.00
	w/582, 684	24.00

Sales Code and UPC	Description	Retail Price ($)
TGF	F78-14 blackwall fiberglass tires (Firebird, Esprit)	
	w/o 582, 684	14.00
	w/o 582, w/684	11.20
THB	E78-14 tires w/white lettering (Firebird, Esprit)	
	w/o 582, 684	28.00
	w/684	22.40
TML	F70-14 blackwall tires w/white lettering (Firebird, Esprit)	
	w/o 582, 684	75.00
	w/582, w/o 684	61.00
	w/o 582, w/684	60.00
	w/684, 582	48.80
TVH	GR-70x15 steel radial tires (Formula)	
	w/684	40.00
	w/o 684	50.00
TVK	GR70-15 steel radial with white lettering Firebird, Esprit	
	w/o 684	92.00
	w/684	73.60
	Trans Am	
	w/o 684	42.00
	w/684	33.60

*Includes FR78-14 whitewall steel radial tires, rear stabilizer bar, special springs, firm shocks, and RTS identification.
**Includes same as above except steel radial tires have raised white lettering.

★	Firebird Six, V-8
★	Esprit Six, V-8
★★	Formula 350
★★	Formula 400
★★★	Trans Am 400
★★★	Trans Am 455

The Firebird's popularity in the showroom continued to increase, even though the energy crunch and government meddling combined to hurt the series' high-performance potential. While the number of models remained unchanged, the options list was pared down initially. In the middle of the model run, the regular 455-ci V-8 was reinstated after having been dropped in the fall.

Radial tuned suspension (RTS), radial tires, and electronic ignition were standard for all Firebirds including this base coupe. This new grille treatment was decidedly more refined; less aggressive than earlier Firebirds, and not nearly as popular. *Pontiac Motor Division*

The 1975's biggest change in styling was a new wrap-around backlight (rear window). The foam-padded front and rear bumper systems were beefed up. At the front of all models, the grille had a new appearance with horizontal blades and square running lamps tucked into each outer corner. Interior alterations included a new speedometer (marked in miles and kilometers) and the discontinuance of the corduroy-cloth upholstery option.

Standard tire equipment on all Firebirds was upgraded to 15-inch steel-belted radials—GR70-15 on the Trans Am and FR78-15 on other models. The Radial Tuned Suspension (RTS) was included.

Another new-for-1975 feature was a demand-type windshield washer system. High-Energy Ignition (HEI) was used with all engines, along with catalytic converters that dictated the use of low-octane, unleaded fuel.

The base Firebird, listing for $3,713, featured "moon" hubcaps and a deluxe two-spoke steering wheel as standard equipment. Compression on the six was boosted very slightly to 8.25:1, giving 105 net horsepower at 3,800 rpm. Torque production was 185 foot-pounds at 1,200 rpm with the six and 280 foot-pounds at 2,000 rpm with the V-8.

With a base price of $3,958, the Esprit came standard with the 250-ci six. Other distinctions of this model were the same as in the previous season. The 350-ci V-8 could be added at extra cost.

The Firebird Formula had slightly wider

tracks, front and rear, than any of the other models. Its retail price started at $4,349 for 350 and 400 models. In this car line, the 350 was the four-barrel version, listed at 20 net horsepower higher than the two-barrel version, but the torque rating for both was the same.

Color changes were about the only alterations that Trans Am fans could brag about in 1975. New finish colors included sterling silver and stellar blue (in place of admiralty blue). The Trans Am also continued to be available in exterior colors of cameo white and buccaneer red.

Otherwise, prices were up and performance was down considerably, which seemed like a double blow to enthusiast buyers. With a base retail price of $4,740 (up almost $300) the top Firebird came with only one engine at the start of the year: the 400-ci V-8 with a new 7.6:1 compression ratio. Specifications for it were 185 net horsepower at 3,600 rpm and 310 foot-pounds of torque at 1,600 rpm. In addition, higher (numerically lower) axle ratios were used, resulting in an even greater performance drop.

Exterior and interior features of the Trans Am package were carried over from 1974 with no major changes except for the new grille and backlight treatments. Engine call-outs on the side of

the shaker hood scoop were initially limited to one choice. The decals, placed near the rear of the scoop, read 400.

This engine limitation raised complaints from performance enthusiasts and threatened to hold back some sales. Management decided to add an optional powerplant, announcing the change on April 25 for cars being assembled in May and later. One reason for the switch may have been Chevrolet's dropping of the Camaro Z-28. A new engine seemed to be one way to draw pony car buyers away from Chevy dealerships and into Pontiac showrooms. Getting a few customers to change brands was more important than making a few extra sales of Firebirds, since Pontiac's F-cars were already selling well.

Labeled the 455 HO, the new option was nothing more than PMD's standard passenger-car engine (455-ci) shoe-horned into the Pontiac F-car. It was not a true high output engine in the sense of previous Pontiac HOs, but its 200 net horsepower was still powerful in comparison to the available offerings from other automakers in 1975.

The 455 HO package also included a four-barrel carburetor, semi-metallic brake linings, four-speed transmission, 3.23:1 rear axle ratio,

In profile, the new 1975 rear window contour is readily apparent on this base coupe with optional finned wheel covers, white-stripe tires, dual sport mirrors, and 350-ci (note fender call-outs) V-8 engine. *Pontiac Motor Division*

The lack of front fender call-outs on this Esprit indicates that it has the six. The delicate, twin pinstripes on the fenders and door were an extra-cost option. The Rally IIs and white-stripe tires were again optional. *Pontiac Motor Division*

and tuned exhaust system with dual splits located behind the catalytic converter. Buyers in California were not allowed to have even this watered-down big-block, however.

Specifications for the midyear option included a 7.6:1 compression ratio. Peak horsepower was developed at 3,500 rpm and the torque rating was a mild 350 foot-pounds at 2,000 rpm.

Pontiac retained its basic coding system from 1974, using a few new VIN alphabetical characters to designate new engines, and changing the sixth character to a 5 to indicate model year 1975. The codes for engines were as follows:

Cars identified with this script continued to come standard with wider rocker panel moldings, dual mirrors, door handle inserts and Custom interior, deluxe wheel covers and twin horns.

Type	Displacement	Carburetor	Exhaust	Code
L-6	250 ci	1V	single	D
V-8	350 ci	2V	single	M
		4V	dual	H
	400 ci	4V	dual	R, S
	455 ci	4V	dual	W

Two-character codes stamped on the block on Pontiac-built powerplants indicated the type of engine in the cars and the factory transmission attachments. For 1975, the codes were:

Engine	Displacement	Compression	Horsepower	Transmission	Code
V-8	350 ci	7.6:1	155	manual	—
				automatic	YA, YB
			175	manual	WN
				automatic	YV, ZP
	400 ci	7.6:1	185	manual	WT
				automatic	YS, YM, YT, ZT
	455 ci	7.6:1	200	manual	WX
				automatic	—

Total Firebird sales for 1975 increased by over 10,000 units although deliveries of base, Esprit, and Formula models all declined. Trans Am sales and production increased approximately 250 percent, offsetting the temporary drop in popularity of the three other models. Available production totals and breakouts were as follows:

Name	Engine	Code	Manual	Automatic	Total
Base		FS87	—	—	22,293*
Esprit		FT87	—	—	20,826*
Formula		FU87	—	—	13,670
Trans Am	400 ci V-8	FV87	6,140	20,277	26,417
Trans Am	455 ci V-8	FV87	857	—	857
Total					84,063**

*Base and Esprit totals include 8,314 cars with six-cylinder engine.

**Includes 70,941 cars with automatic transmission, 10,775 cars with four-speed manual transmission, and 2,347 cars with three-speed manual transmission.

Reliability and repair information that potential buyers and collectors of 1974 Firebirds might want to see is sparse. A generalized look at these models is offered in *Consumer Guides Used Car Rating and Price Guide 1971–1980*. It indicates an overall repair rate that was slightly worse than average.

Other Firebird experts suggest that 1975's performance cuts were somewhat offset by improvements in maintenance and the new electronic

There was also an Esprit script on the right-hand corner of the trunk. The new wrap-around rear window gave a better view of the road behind the Firebirds. The wider taillights and rubber bumper, introduced a year earlier, were carried over for 1975 models. *Pontiac Motor Division*

ignition system. For example, Joe Oldham says that the HEI system "greatly reduced maintenance and helped engine performance".

Gary Witzenberg notes of the catalytic converter, "Its principal advantage was that engines could then be calibrated on the 'dirtier'—and therefore better running and more fuel efficient—side, and the converter would clean up the carbon monoxide (CO) and hydrocarbon (HC) pollutants sufficiently to meet emissions standards."

In regard to HEI, Witzenberg states the cars

The Formula didn't look quite as mean with its 1975 front end, although the double forward scoops were seen again. No longer available was the Formula 455. Still around, however, were the Formula 350 and the Formula 400 (seen here). A four-speed transmission, front disc brakes, and front and rear stabilizer bars were standard on all Formulas.

The Custom interior option was offered in Saddle, burgundy, white, blue, and black for 1975. Also available were blue, black, burgundy, and saddle interiors with white seats and five additional "special appointment" combinations. Materials in the Custom option were called perforated and Madrid Morrokides (Morrokide vinyl).

The standard interior came in black, white, or saddle regular combinations; black or saddle colors with white seats; or in four special combinations. It was all done in Madrid Morrokide in patterns similar to previous years. All corduroy and cloth Custom options were dropped for 1975.

Here's a 1975 Firebird Esprit coupe without the optional pinstriping. The only option on the car seems to be white-stripe tires. They look quite attractive in combination with the regular deluxe wheel covers.

gained "better drivability and some fuel economy improvements over '73-'74 models" and "the added benefit of longer oil, oil filter, chassis lube, and other service intervals."

The HEI system also more-than-tripled spark plug life; recommended intervals for plug changes were upped from 6,000 to 22,500 miles.

According to published figures, fuel economy ratings for the 1975 Firebird engines were not noticeably higher. The six was rated for up to 17.5 miles per gallon; the 350 two-barrel, for 16 miles per gallon; the 350 and 400 four-barrels, for 15 miles per gallon; and the new 455, for about 13.5 miles per gallon tops.

In summary, collectors interested in performance only are probably better off shopping for pre-1975 Firebirds. On the other hand, those looking for a car mainly for regular driving may prefer 1975 or newer versions for their longer service intervals.

The National Highway Traffic Safety Administration issued two recalls covering problems with 1975 Firebirds, and collectors will want to make sure that these problems do not exist in cars they are considering. The first recall was made to check the torque on the cars' steering post joint clamp bolts or nuts. These fasteners secured the intermediate steering shaft to the upper steering shaft. Insufficient torque could eventually lead to

The 455 HO engine reissued in 1975 was basically the same powerplant that was standard on the Catalina Safari and Grand Safari station wagons. It featured a cast-iron block, five mains, aluminum pistons, integral valve guides, fully machined combustion chambers, rolled crankshaft fillets, high-capacity oil pump, large distributor drive gear, swirl-polished valves, and a Quadrajet carburetor.

loss of steering control.

The second recall applied only to Trans Ams with the 455-ci engine and instructed dealers to check the brake system and brake combination valve. Some cars had standard brake-pad linings, instead of semi-metallic linings.

1975 Firebird Options and Accessories

Sales Code and UPC	Description	Retail Price ($)
36D/L22	250-ci 6-cyl. (S87/T87)	n./c
36M/L30	350-ci V-8 2V (S87, T87)	$ 130.00
36E/L76	350-ci V-8 4V	
	S87, T87	180.00
	U87	n/c
36S/L78	400-ci V-8 4V	
	U87	56.00
	V87	n/c
—/L75	455-ci V-8 4V (V87)	150.00
37L/M38	Turbo Hydra-matic trans.	
	S87, T87	237.00
	U87, V87	n/c
37B/M15	Three-speed man. trans. w/floor shift	
	(S87, T87)	n/c
37E/M20	Four-speed man. trans.	
	S87, T87	219.00
	U87, V87	n/c

Sales Code and UPC	Description	Retail Price ($)
37F/M21	Close-ratio four-speed man. trans.	
	(U87, V87)	n/c
602/K45	H-D air cleaner (std. w/Trans Am)	11.00
582/C60	Custom air conditioning	435.00
381/G80	Safe-T-Track axle; differential	
	V87	n/c
	S87, T87, U87	49.00
591/UA1	Custom seat belts; and Soft-Tone warning	19.00
541/VJ9	Calif. emissions equip. and test	45.00
711/U35	Elec. clock w/o Rally gauges or speedo. and clock	16.00
431/D55	Front seat console	68.00
424/D58	Rear seat console	41.00
594/C50	Rear window defogger	41.00
592/C49	Elec. rear window defroster	70.00
601/K05	Engine block heater	11.00

Sales Code and UPC	Description	Retail Price ($)
704/UR1	Fuel economy vacuum gauge (exc. V87)	22.50
712/W63	Rally gauges and clock (S87, T87, U87)	50.00
714/WW8	Rally cluster, clock and tach. (n/c w/V87)	99.00
571/AQ1	Soft-Ray glass, all windows	43.00
572/AQ2	Soft-Ray glass, windshield only	34.00
492/B93	Door edge guards	7.00
444/T63	Headlamp warning buzzer	6.00
512/WW7	Trans Am hood decal (V87 only)	55.00
584/BS1	Added acoustical insulation (std. in T87)	20.00
654/Y92	Lamp group (glovebox, instrument panel courtesy, luggage lamps)	11.50
622/B32	Front floor mats	8.00
624/B33	Rear floor mats	7.00
422/D35	Sport mirrors; LH remote control, RH fixed	
	S87	27.00
	T87, U87, V87	n/c
432/D34	RH visor vanity mirror	3.00
494/B84	Bodyside moldings (except V87)	35.00
481/B80	Roof drip moldings	
	std. w/T87, w/o Cordova vinyl trim	15.00
	w/Cordova vinyl top	n/c
491/B96	Wheel lip moldings (S87, U87; std. w/T87; n.a. w/V87)	16.00
484/B85	Windowsill and rear hood edge molding (std. W/T87)	21.00
502/JL2	Front power disc brakes (std. w/U87, V87)	21.00
554/AU3	Power door locks	56.00
551/A31	Power windows	91.00
401/U63	AM radio	69.00
403/U69	AM/FM radio	135.00
405/U58	AM/FM stereo radio	233.00
684/N65	Space-Saver spare	n/c
804/U18	Speedometer w/kilometers & clock	21.00
411/U80	Rear seat speaker	19.00
681/D80	Rear air spoiler (na T87; std. w/V87)	45.00
461/D80	Custom cushion steering wheel (S87 only)	16.00
464/NK3	Formula steering wheel	
	S87	57.00
	Std. w/V87, T87, U87	41.00
504/N33	Tilt steering wheel	49.00
412/U57	Stereo 8-track player	130.00
638/D98	Accent stripes (exc. V87)	43.00
PVT/CO9	Cordova vinyl top	99.00
32W/Y90	Custom trim group (U87, V87)	81.00
511/JL1	Pedal trim pkg. (std. w/T87, Y90 option)	5.00

Sales Code and UPC	Description	Retail Price ($)
472/PO2	Custom finned wheel covers	
	S87, U87	54.00
	T87	24.00
476/PO1	Deluxe wheel covers (n/c w/T87, S87, U87)	30.00
478/P05	Honeycomb wheels (n/c w/V87)	
	S87, U87 w/o 684	127.00
	S87, U87 w/684	107.60
	T87 w/o 684	97.00
	T87 w/684	77.60
474/N98	Rally II wheels & trim rings (n/c on V87)	
	S87, U87 w/o 684	91.00
	S87, U87 w/684	78.80
	T87 w/o 684	61.00
	T87 w/684	48.80
471/P06	Wheel trim rings (S87, U87 w/o 474)	30.00
414/C24	Recessed wipers (S87, U87)	18.00
39A/QBU	FR78x15 blackwall steel-belted radial tires (Firebird, Esprit, Formula)	n/c
39B/QBX	GR70x15 blackwall steel-belted radial tires (Formula)	
	w/o Space-Saver spare	40.00
	w/Space-Saver spare	34.40
39L/QBP	FR78x15 white-letter steel-belted radial tires (Firebird, Esprit, Formula)	
	w/o Space-Saver spare	43.00
	w/Space-Saver spare	34.40
39M/QXCY	GR70x15 white-letter steel-belted radial tires	
	Formula	
	w/o Space-Saver spare	85.00
	w/Space-Saver spare	68.00
	Trans Am	
	w/o Space-Saver spare	45.00
	w/Space-Saver spare	36.00
39W/QBW	FR78x15 whitewall steel-belted radial tires (Firebird, Esprit, Formula)	
	w/o Space-Saver spare	33.00
	w/Space-Saver spare	26.40
39Y/QFM	F78x14 whitewall fiberglass tires (Firebird, Esprit)	
	w/o Space-Saver spare (credit)	(73.00)
	w/Space-Saver spare	(60.40)
39Z/QFL	F78x14 blackwall fiberglass tires (Firebird, Esprit)	
	w/o Space-Saver spare	(106.00)
	w/Space-Saver spare	(86.80)

Chapter 10

1976

★	Firebird Six, V-8
★	Esprit Six, V-8
★★	Formula 350
★★★	Formula 400
★★★	Trans Am 400
★★★	Trans Am 455
★★★★	Trans AM 400 Limited Edition
★★★★★	Trans Am 455 Limited Edition

New front and rear styling that featured integrated, body-color urethane bumpers made 1976 Firebirds the most-changed in five years. Styling elements in front had a slightly more rectangular look, with smaller, narrower air inlets in the lower valence. The parking lights were moved into the outboard corners of these slotlike openings below the bumper.

Brake systems on all Firebirds were refined to meet changed federal standards, axle ratios were lowered (for better economy), V-8s idled a little slower, and there were cooling system and air-conditioning system improvements. The Radial Tuned Suspension (RTS) no longer incorporated a rear sway bar.

Seat patterning in the standard interior was converted to vertical ribbing, with seatbacks having three distinct sections. The door panels had twin, vertical-ribbed sections that were similar but not identical to the 1975 style. The Custom interior featured wider vertical ribs. On the seatbacks the ribs were bordered by a smooth "horse collar." On the seat cushions, the ribs ran fully over the seat and down to the floor. Custom door panels were also similar to the previous year's design.

Engines were basically those of 1975, but with some horsepower adjustments and a slightly higher 8.3:1 compression for the six cylinder. Manual transmissions and some engines were not available for cars sold in California.

There were some color changes in both exterior finish and upholstery. An interesting new "appointment mix" combination was white with lime (green). The Formula got a lot of attention,

with a new Formula appearance package adding to its excitement level. A Limited Edition Trans Am appeared at midyear. Total Firebird sales exceeded 100,000 units for the first time.

The base Firebird, listing for $3,905.80, had about the same standard equipment as in 1975. A Rosewood instrument panel appliqué replaced the crossfire mahogany insert. A new brake-fluid sensor was added. Turbo Hydra-matic became a required option in California and a 3.08:1 rear axle was regular fare. Narrow rocker panel moldings and windshield and rear window moldings were included. A canopy-style vinyl half-top was a new option for all 1976 Firebirds.

The standard version of the new seat patterning, with the four-speed shifter, is shown in this 1976 Firebird.

Horsepower for the six was up to 110 at 3,600 rpm. Options included a two-barrel 350-ci V-8 (not available in California), a four-barrel 350-ci V-8 (California only), and a four-barrel 400-ci V-8 (nationwide). The 400 was a new option for the base model and the Esprit. Horsepower ratings for the three powerplants were 160 net horsepower at 4,000 rpm, 165 net horsepower at 4,000 rpm, and 185 net horsepower at 3,600 rpm, respectively.

Esprits came with the normal additional features inside and out. Base price was $4,161.80 Brightening the exterior were wide rocker panel moldings and bright metal trim on the window sills, rear edge of hood, rain gutters, and wheel lips. Esprit scripts were again carried on the front fender side with chrome call-outs for optional V-8s below them.

Engine and transmission options for the Esprit were the same as for base Firebirds at similar prices. Both also used the 3.08:1 axle and FR78-15 blackwall steel-belted radial tires. As usual, the Custom trim option was standard in Esprits.

While the base Firebird's price climbed $180 and the Esprit's was up $191, the increase for 1976 Formulas was $204, suggesting a few more equipment additions. Among new items found on the $4,565.80 Formula were an all-steel simulated air scoop hood, a front seat console, and chrome side-splitter type tail pipe extensions. It was the first time a console—a full-length design—was standard equipment in any Firebird.

The base Formula engine was switched to the two-barrel version of the 250, which came only in federal marketing areas and only with the Turbo Hydra-matic. This powerplant was not legal in California. Buyers in that state were required to pay $55 extra to have the 350-ci four-barrel V-8 as a mandatory option. It, too, came only with a Turbo Hydra-matic transmission. (Technically, there was no base engine in California.)

The 400-ci V-8 was available nationwide at $118 extra. It could be had with the four-speed manual transmission as the standard transmission outside California. Turbo Hydra-matic was a no-cost option and was required in California. The automatic came only with floor-mounted controls.

The same exterior moldings, hubcaps, tires, and interiors that were regular equipment with 1976 Formulas were included for base Firebirds. So the Formula was essentially a plain-Jane high-performance machine. New, however, was the optional Formula appearance package (code W50). Cars with the extra had the whole lower

The 1976 Esprit had Firebird emblems on the roof sail panels, instead of signature-type scripts. Newly available on all Firebird models was this canopy roof treatment; the vinyl front section came in colors of white, black, buckskin, mahogany, firethorn, blue, and silver. This type of grille with hex-pattern screening and redesigned front bumper gave a lighter, more agile look. *Pontiac Motor Division*

perimeter of the body finished in a contrasting color with multicolor striping above. The same motif was carried to the air scoops and Formula graphics were emblazoned on the lower sides of the body.

At first, the W50 package came only with yellow as the main body-finish color and black-out finished lower body perimeter. However, three other color combinations were added later. The four color combinations carried codes C, N, T, and V. Formulas with the $100 option proved to be very popular and soon accounted for over 85 percent of Formula sales.

Priced at $4,987.22, the 1976 Trans Am was up $234.12 in base price, although its list of standard extras was virtually the same as in 1975.

The base engine used in the Trans Am, throughout the nation, was the 400-ci V-8 with four-barrel carburetor—the same engine that could be added to base Firebirds. As with the Formula, a four-speed manual transmission was standard, but not available in California, and a Turbo Hydra-matic was a no-cost option. For $125 more, buyers outside California could add the 455-ci V-8. It, too, rated 200 net horsepower at 3,500 rpm.

The 455 was for the Trans Am only and was merchandised as a performance package including specific identification, specifically tuned exhaust, 3.23:1 rear axle, and close-ratio four-speed manual transmission. Since the four-speed was not sold in California, the 455 wasn't available there, either.

Introduced at the Chicago Automobile Show in February 1976 was the special Limited Edition Trans Am. Featuring a starlight black finish and gold striping, gold interior accents, gold honeycomb wheels, and a host of other distinctions, the car was promoted as a commemorative model issued in honor of the Pontiac nameplate's 50th anniversary.

This car was a new Trans Am with a conversion added by the Hurst Corporation of Michigan. The overall character of the car was based on a 1974 show car designed by stylist John Schinella and inspired by the black-and-gold John Player race cars. Part of the package was a gold, over-sized version of Schinella's famous "screaming chicken" Trans Am hood decal.

Production of Limited Edition models was originally projected at 2,500 units, but there were some manufacturing problems involving the

Formulas were treated to a major facelift and new duct-type twin-scoop hood. The model name appeared below the Firebird nameplate on the front fenders. The Formula was beginning to look more like a sports car and less like a hopped-up street machine. But those fat, white-letter radial tires still meant business.

Hurst hatch roof treatment seen on the original show car and planned as a production option. This T-top treatment (a hatch roof with removable panels) was originally created for the 1976 Pontiac Grand Prix and its adaptation to the Firebird body didn't go smoothly. Consequently, production of the Limited Edition did not start until April and, even then, the Hurst hatch was used only on about 25 percent of the cars.

Total production turned out to be 2,590 units (not the 2,400 that some books say), and only 643 had T-tops. Also relatively rare was installation of the optional 455-ci engine, which went into less than 500 Limited Edition Trans Ams. However, all of these cars had a few other common features, such as gold instrument panel appliqués, gold-spoked steering wheels, gold grilles and headlight liners, and gold emblems inside and out.

As a collectible automobile, the Limited Edition model has a lot of appeal to knowledgeable Pontiac enthusiasts because it was the prototype for the similar-looking Special Editions that came later. The car to go for is the T-top version with the 455-ci engine, of which only 110 were assembled.

Pontiac used similar VIN codes in 1976,

changing the sixth character to a 6 to announce the new model year. The S, T, U, and W symbols were again used to designate Firebird, Esprit, Formula, and Trans Am car lines, respectively. Each was combined with the hardtop body code 87 to form the model number (for example, S87 for the base Firebird). In some sources, the number might also appear as 2FS87, the 2 designating Pontiac Motor Division and the F standing for the F-body GM vehicle platform. The new Limited Edition without T-tops was designated Y81, and Y82 was used as the model code when the optional roof was added. Alphabetical codes for different engines were as follows:

Engine	Displacement	Code
L6	250 ci	D
V-8	350 ci	M, J, H, P
	400 ci	N, Z
	455 ci	W

Two-character codes stamped on the block on Pontiac-built powerplants designated the engines and type of transmission they were teamed with. The 1976 codes were:

Engine	Displacement	Compression	Horsepower	Transmission	Code
V-8	350 ci	7.6:1	160*	manual	—
				automatic	YA, YV, YP, YR
			160**	manual	—
				automatic	YM, XP, XU, XR, XW, XX, X3, ZF, ZH
			165	manual	—
				automatic	ZX, ZD
	400 ci	7.6:1	170*	manual	—
				automatic	XJ, XC
			170**	manual	—
				automatic	X9, X4, XT, XY, X5, XZ, X8
			185*	manual	WT, WU
				automatic	Y6, Y7, YS, YT, YY, YZ, ZA, ZK
			185**	manual	WY
				automatic	X7, ZJ, ZL, X6
	455 ci	7.6:1	200	manual	WX
				automatic	Y3, Y4, Z3, Z4, Y8, Z6

Total Firebird production for 1976 was up for Esprits, Formulas, and Trans Ams—especially Trans Ams. About 1,000 fewer base Firebirds were built. This was indicative of the way things were moving upscale in what remained of the pony car marketplace. The Esprit was, for a time, a modest success, while the Trans Am was a clear winner. But base Firebirds and Formulas never really got hot; they didn't have the luxury image. Available production totals and equipment breakouts for 1976 looked like this:

Name	Engine	Code	Manual	Auto.	Total
Base Firebird		S87	—	—	21,106
Esprit		T87	—	—	22,252
Formula		U87	—	—	20,613
Trans Am	400 ci V-8	W87	—	—	37,015
	455 ci V-8	W87	7,099	—	7,099
Limited Edition Coupe	400 ci V-8	Y81	—	—	1,628
	455 ci V-8	Y81	319	—	319
Limited Edition T-top	400 ci V-8	Y81	—	—	533
	455 ci V-8	Y82	110	—	110
Total					110,775

According to owner surveys, the 1976 Firebirds were still less reliable than the models of the early seventies. They had an overall repair rating that was slightly worse than average for American cars. Most prominent among owner complaints were three items: the tendency to rust-out rather quickly; poor-quality paint; and loose, rattle-prone body construction.

There were two factory recalls, both on relatively dangerous conditions. On cars with 400-ci engines, dealers were asked to check the fuel vapor return pipe. Too short a pipe could cause fuel to spill and possibly cause a fire in a crash. The second recall was to check the power brake booster and brake diaphragm. Under certain conditions, gasoline vapors could condense in the power brake booster. If liquid gasoline contacted the rubber brake diaphragm, it could cause enough deterioration to cause a rupture, resulting in loss of power assist upon brake application.

For their performance, the 1976 models received praise from several magazines. Road Test took a liking to the 455 Trans Am in its September 1975 issue. *Cars* named the same model Top Performance Car of the Year. Joe Oldham was *Cars'* editor at the time and found the 1976 model superior to its 1975 counterpart. He also considered it the "best looking" of the all Trans Ams.

In April 1976, *Car and Driver* sought out America's fastest car. The 455 Trans Am finished second to an L82 Corvette. The Pontiac's top speed was 117.6 miles per hour. It ran 0–60 miles per hour in seven seconds and did the quarter-mile in 15.6 seconds at 90.3 miles per hour.

The 1976 model year holds some special interest to collectors because it was the first season for the Formula Appearance package, the last year for the revived 455-ci V-8, and the start of

Pontiac maintained its leadership in the 1976 sports car market with the Trans Am coupe. The shaker hood, side spoilers, rear decklid spoiler, front air extractors, Rally gauge cluster, and other well-known goodies were standard equipment again.

Highly promoted in 1976 advertising was the Formula in Goldenrod yellow with black lowerbody perimeter color. A red line running around the car, between the two hues, was complemented with red Formula lettering above the outboard corner of the left-hand grille. Blacked-out finish, again with red pinstriping, set off the forward edge of the twin hood-scoop openings. The shadow lettering on the rocker panels was yellow, as was the spyder section behind the chrome-spoked Rally II wheels. The grille was blacked-out, too.

Released as a Pontiac Limited Edition model for the company's 50th anniversary in 1976 was this special black and gold Trans Am Limited Edition. Its many standard features and options included custom trim and a T-top with removable hatch panels. This Firebird factory custom was based on a 1974 concept car. There were 1,947 black-and-gold Limited Edition coupes (without T-tops) made. *Pontiac Motor Division*

the Trans Am Edition options. The black-and-gold 50th Anniversary Limited Edition was virtually the prototype for the black-and-gold Special Edition package, which remained very popular among enthusiasts for several years (it was joined later by gold Special Editions and Daytona and Indy 500 Pace Cars).

Being the first in the series makes the Limited Edition special to start with. In addition, production was relatively low. Then too, it was the only black-and-gold special that came with the 455 Trans Am option, which makes it that much more exciting to the serious collector. The car didn't have the performance of earlier Trans Ams nor the luxury level of later ones, but it is definitely historically significant, unique, and well-balanced overall. And since it is a high demand, low-availability car, it has been given the top, five-star rating in 455 trim.

As for other 1976 models—Firebirds, Esprits, and regular Formulas—collector interest levels range from poor to good. The best investment of the year may well be the Limited Edition with the base 400-ci engine, which isn't as exotic nor pricey as the 455, but still has a lot of visual appeal.

1976 Firebird options and accessories

Sales Code and UPC	Description	Retail Price ($)
37D/L22	250-ci 6-cyl. 1V (S87/T87)	n/c
37M/L30	350-ci V-8 2V (S87, T87; na Calif; std. w/U87; req. w/Turbo Hydra-matic	140.00
37E/L76	350-ci V-8 4V S87, T87; Calif. only; Req. w/Turbo Hydra-matic	195.00
	U87; Calif. only, Req. w/Turbo Hydra-matic	55.00
37S/L78	400-ci V-8 4V S87, T87; n.a. Calif; req. W/Turbo Hydra-matic, 38F	258.00
	U87; n.a. Calif; req. Turbo Hydra-matic, 38F; std. w/W87	118.00
37W/L75	455-ci V-8 4V (W87 only; na Calif; req. w/Turbo Hydra-matic, 38F)	125.00
38B/M15	3-speed man. trans. (incl. floor shift; w/37D only)	n/c
38F/M21	4-speed man. close ratio; in W87; n.a. Calif.	242.00
	close-ratio; in W87; n.a. Calif.	n/c
38L/M40	Turbo Hydra-matic (S87, T87; std. w/U87, W87)	262.00
492/C60	Custom air conditioning	452.00
674/K97	H-D alternator V-8s only; w/o C50, C50; w/C60	42.00
	w/C50, C60	n/c
582/W50	Formula Appearance package (colors C, N, T, V only)	100.00

Sales Code and UPC	Description	Retail Price ($)
391/G80	Safe-T-Track axle (S87, T87, U87; std. w/W87)	51.00
681/UA1	H-D battery	16.00
524/AK1	Custom front & rear seat belts & front shoulder belt	20.00
721/VJ9	Calif. emissions equipment & test	50.00
474/U35	Elec. clock (n/c w/W63, WW8)	18.00
581/D55	Front console (req. w/Turbo Hydra-matic; std. w/U87)	71.00
572/D58	Rear seat console	43.00
462/C50	Rear window defogger	48.00
461/C49	Elec. rear window defroster (V-8s only)	17.00
684/K05	Engine block heater	12.00
481/UR1	Vacuum gauge (S87, T87, U87 w/W63 only)	25.00
502/W63	Rally gauge cluster & clock (S87, T87, U87; req. w/UR1 w/six)	54.00
504/WW8	Rally gauge cluster, w/clock & tach. (S87, T87, U87, V-8 only)	106.00
442/A01	Soft-Ray tinted glass, all windows	46.00
612/B93	Door edge guards.	8.00
664/WW7	Hood decal (W87 only)	58.00
521/BS1	Added acoustical insulation (std. w/Esprit)	25.00
631/Y92	Lamp group (glovebox, instrument panel courtesy, luggage comp.)	14.00
601/B32	Front floor mats	8.00
602/B33	Rear floor mats	7.00
642/D35	Dual sport OSRV mirror, LH remote control (std. w/T87, U87, W87)	29.00

Sales Code and UPC	Description	Retail Price ($)
652/D34	Visor vanity mirror	4.00
611/B84	Color-keyed vinyl body moldings (S87, T87, U87)	38.00
584/B80	Roof drip moldings (S87, U87 w/o CB7; W87; std. w/T87, CB7)	16.00
614/B96	Wheel lip moldings (S87, U87; std. w/T87; n.a. w/W87)	17.00
591/B85	Window and rear hood moldings (S87, U87, W87; std. w/T87)	22.00
452/JL2	Power front disc brakes (req. w/V-8; std. W/U87, W87)	58.00
434/AU3	Power door locks	62.00
431/A31	Power windows (w/D55 only)	99.00
682/VQ2	Super cooling radiator (w/o C60; $22 less w/C60)	49.00
411/U63	AM radio	137.00
415/U69	AM/FM radio	137.00
415/U58	AM/FM stereo radio	233.00
802/UN9	Radio accommodation pkg. (n/c w/optional radios)	22.00
441/N65	Stowaway spare (S87, T87, U87; std. w/W87)	(1.22)
421/U80	Rear seat speaker (optional only w/U63, U69)	20.00
562/D80	Rear air spoiler (S87, T87, U87; std. w/W878)	48.00
541/N30	Custom cushion steering wheel (S87; std. w/T87, U87)	17.00
544/NK3	Formula steering wheel	
	S87; n/c w/W87	60.00
	T87, U87; n/c w/W87	43.00
444/N33	Tilt steering wheel	52.00
422/U57	Stereo 8-track player (req. W/D55 & radio)	134.00
358/D98	Vinyl accent stripes (S87, T87, U87)	46.00
CVT/CB7	Canopy top (S87, T87, U87; n.a. w/W87)	96.00
32N/Y90	Custom trim group (U87, W87 only; std. w/T87)	81.00
561/JL1	Pedal trim pkg. (S87, U87, W87 w/o Y90; std. W/T87)	6.00
556/P01	Deluxe wheel covers (S87, U87; n/c w/T87; n.a. w/W87)	6.00
558/P05	Honeycomb wheels, four, w/radials (S87, U87)	135.00
558/P05	Honeycomb wheels, four, w/radials (T87; n/c w/W87)	103.00

Sales Code and UPC	Description	Retail Price ($)
554/N98	Argent silver Rally II wheels & trim rings	
	w/fiberglass tires w/o N65 (5 wheels) on S87	113.00
	w/fiberglass tires w/N65 (4 wheels) on S87	97.00
	w/o fiberglass tires (4 wheels) on S87	97.00
	w/fiberglass tires w/o N65 (5 wheels) on T87	81.00
	w/fiberglass tires w/N65 (4 wheels) on T87	65.00
	w/o fiberglass tires (4 wheels) on T87	65.00
	w/o fiberglass tires (4 wheels) on U87	97.00
	w/o fiberglass tires (4 wheels) on W87	n/c
559/N67	Body-color Rally II wheels & trim rings	
	w/fiberglass tires w/o N65 (5 wheels) on S87	113.00
	w/fiberglass tires w/N65 (4 wheels) on S87	97.00
	w/o fiberglass tires (4 wheels) on S87	97.00
	w/fiberglass tires w/o N65 (5 wheels) on T87	81.00
	w/fiberglass tires w/N65 (4 wheels) on T87	65.00
	w/o fiberglass tires (4 wheels) on T87	65.00
	w/o fiberglass tires (4 wheels) on U87	97.00
	w/o fiberglass tires (4 wheels) on W87	n/c
574/C24	Recessed windshield wipers (S87, T87; std. w/U87, W87)	22.00
40B/QBU	FR78x15 black-belted radial tires (S87, T87, U87)	n/c
40W/QBW	FR78x15 whitewall steel belted radials	
	w/o N65; exc. W87	35.00
	w/N65; exc. W87	28.00
40L/QBP	FR78x15 white-letter steel-belted radials	
	exc. W87; w/o N65	46.00
	exc. W87; w/N65	36.80
40G/QBX	GR70x15 blackwall steel-belted radials (U87; std. w/W87)	34.44
40F/QCY	GR70x15 white-letter steel-belted radials	
	U87; req. w/N65	72.84
	W87; req. w/N65	38.40
40D/QFM	F78x14 whitewall fiberglass radials	
	S87, T87; w/o N65	(62.45)
	S87, T87; w/N65	(49.96)
40C/QFL	F78x14 blackwall fiberglass radials	
	S87, T87; w/o N65	(97.45)
	S87, T87; w/N65	(77.96)

Chapter 11

1977

★	Firebird V-6, V-8
★	Esprit V-6, V-8
★★	Esprit Sky Bird
★	Formula
★★	Formula with W50
★★	Trans Am
★★★	Trans Am with Y81
★★★★	Trans Am with Y82

For 1977, Firebirds received another front end revision. It featured dual, rectangular headlights. The grille bottom was extended toward the center, which gave the nose a more aggressive V shape. The Formula and Trans Am also had new hoods.

Appearance packages were available to add distinction to specific models. These are of more interest to collectors than most 1977 engine options. One big change in the powerplant department was the disappointing disappearance of the 455-ci V-8. Only one performance engine remained.

Prices for the base Firebird six rose considerably, to $4,270. This may have been partly because a Buick-built 231-ci V-6 replaced the inline Chevrolet six. With a two-barrel carburetor and 8.01:1 compression, this new engine developed 105 net horsepower at 3,200 rpm and 185 foot-pounds of torque at 2,000 rpm.

The balance of the standard equipment list was pretty much a carryover from 1976, but there were some minor changes. They included a new, three-spoke deluxe cushion steering wheel and a speedometer that displayed both kilometers and miles per hour. Interior color mixes were revised to give new combinations of white vinyl seats with different-color appointments.

Other characteristics of the standard, base Firebird included bright metal door handles, Pontiac lettering in the left-hand grille, narrow rocker panel moldings, and windshield and rear window moldings. The regular, small hubcaps changed from the old, babymoon style to a ridged design with a V-shaped Pontiac crest. Firebird lettering

was carried on the lower front fender, behind the wheel opening. Cars with either of two optional V-8s were identified with 5.0-liter (301-cubic inch) or 5.7-liter (350-cubic inch) badges on the rear. There were actually two 350-ci engines—one was used only in cars sold in California, the other for every place but California.

The V-6 was also standard in the Esprit, which was base priced at $4,551. This 1977 model had some extra attention lavished on it, since the market was moving away from performance and toward luxury.

Deluxe wheel covers were designed. They had three bands of different-sized, round perforations and smaller center inserts. A new, luxury cushion steering wheel was featured. Doeskin vinyl upholstery was a standard part of the Esprit's basic Custom interior. New, extra-cost Lombardy velour upholstery was optional.

Power teams for Esprits were the same as the base Firebirds. With both models, a three-speed manual transmission was standard. A four-speed manual gearbox was available with the 301-ci V-8. An M40 Turbo Hydra-matic could be ordered with any powerplant.

Esprit exterior trim consisted of a bird-shaped badge on the sail panel and an Esprit script below the Firebird lettering on the front fenders. Other standard extras were the same as offered in the past. A canopy top and vinyl accent striping (in specific colors) was optional on all Firebirds except Trans Ams.

Based on a 1976 show car called the Blue Bird was a new W60 Sky Bird option package for Esprits. It featured a two-tone blue finish (medium blue upper, dark blue lower), blue-trimmed snowflake wheels, medium blue grille liners, blue-accented taillight bezels, and Sky

Firethorn was a new color for the standard 1977 Firebird interior. The door panel design was the same as used since 1970 1/2. Patterning on the seats was similar to 1976 style. Shown here is the automatic transmission console with power window controls behind the gearshift lever panel.

Bird sail panel decals. Bodyside tape stripes, in graduating tones of blue, added another special appearance touch.

The Sky Birds had blue Custom interiors, blue Formula steering wheels, blue custom seat belts, and blue velour or doeskin vinyl upholstery. They were essentially direct copies of the show car, but the Blue Bird Body Company of Georgia (best known for its school buses) refused to allow use of the original name. The number of Sky Birds built isn't available, but the package has some extra appeal to modern Firebird collectors and seems to be getting scarce.

Available only with V-8 powerplants, the 1977 Formula had a new hood and dual scoop design, standard rear spoiler, blacked-out grille and chrome side-splitter tail pipes. Other standard equipment in this model included the redesigned, small hubcaps, the new luxury cushion steering wheel, and Formula identification.

The standard engine in Formulas was the Pontiac-built 5.0-liter, two-barrel powerplant. It had an 8.2:1 compression ratio for 135 net horsepower at 4,000 rpm and 235 foot-pounds of torque at 2,000 rpm. A four-speed manual or Turbo Hydra-matic transmission was standard at the base price of $4,977.

The base Firebird coupe (left) for 1977 had new standard features such as a V-6 engine and redesigned wheel covers. The car shown has deluxe wheel covers and the optional bodyside moldings. Showing off its new front end is a 1977 Esprit (right) with the Sky Bird Appearance package, which is of slight collector interest.

Formula engine options included two 350-ci V-8s, two 400-ci V-8s, and a 403-ci V-8 offered with the same transmission choices as the standard 301-ci block. The following specifications are for the optional powerplants:

Code	Source	Certified	Bore x Stroke	Displacement	Carb.	Compr.	Net HP	Torque
L76	Pontiac	federal	3.88x3.75	350 ci (5.7 liters)	4V	7.6:1	170 @ 4,000	280 @ 1,800
L34	Oldsmobile	California	4.06x3.39	350 ci (5.7 liters)	4V	8.0:1	170 @ 4,000	275 @ 2,000
L78	Pontiac	federal	4.12x3.75	400 ci (6.6 liters)	4V	7.6:1	180 @ 3,600	325 @ 1,600
W72	Pontiac	federal	4.12x3.75	400 ci (6.6 liters**)	4V	8.0:1	200 @ 3,600	325 @ 2,400*
L80	Oldsmobile	California	4.35x3.39	403 ci (6.6 liters***)	4V	8.0:1	185 @ 3,600	320 @ 2,200

*Torque w/manual transmission is listed; torque w/Turbo Hydra-matic was 325 @ 2,200.
**Trans Am
***Mandatory in California

A special Formula Appearance package (W50) was again a moderately priced option for Formulas. It included specific lower-body perimeter finish, multicolor stripes, blacked-out simulated hood scoops, and large Formula graphics on the lower body sides.

This package came in six color choices: cameo white with black lower perimeter and blue stripping; sterling silver with charcoal and light red; black with gold; golden red yellow with gold, orange, and black; glacier blue with medium blue and black; and buccaneer red with black perimeter and light and dark red stripes.

Listing for $5,456, the Trans Am was again the ultimate Firebird. Equipment distinctions of the model were essentially the same as before. The new hood had a broader bulge. The redesigned shaker-type hood scoop had a broader

Identification for the 1977 Esprit was provided by chrome signatures spelling out the model name under the Firebird lettering on front fenders. Dual sport-type outside rearview mirrors were standard again. As in the past, the left-hand mirror had the remote-control feature.

look, too, with a crease up its center.

The L78 was the standard powerplant. W72 was the only performance option. It had different cylinder heads and valve timing. The heads were those used on the Pontiac 350-ci engine. Intake valves opened faster and closed slower (5 to 13 degrees in both cases) than those in the base L78. The exhaust valves opened 2 to 17 degrees sooner.

Performance of a Trans Am with the W72 400 engine was recorded by Car and Driver in April 1977. With a top speed of 110 miles per hour, the four-speed equipped Trans Am 6.6 went 0–60 miles per hour in 9.3 seconds and did the quarter-mile in 16.9 seconds at 82 miles per hour. These times were down from those of the top (455-ci) 1976 optional powerplant.

Those who purchased 1977 Trans Ams in California or designated high-altitude counties had to make do with even less excitement. In such regions, the only engine offered was the L80 option with mandatory automatic. This transmission was also mandatory with the L78 block, while the W72 could be had with the M20 wide-ratio four-speed manual gearbox instead.

Like other 1977 Firebirds, the Trans Am could be had with an extra-cost Special Edition package

Grab even more excitement of steering this stylish over the roof stripe,

Firebird's standard instrument panel is dashingly handsome. Rally gauges and the popular tilt steering wheel are available.

Rebirth of the Blues! Now available on Esprit is a special luxury appearance package called Sky Bird. It includes sumptuous blue velour bucket seats, two tone blue exterior, blue cast aluminum wheels, dark blue rear panel, blue grille panels accent stripes, special identification & more.

Firebird's available AM/FM stereo radio. You can even order a separate 8 track tape deck unit.

The 1977 Firebird catalog showed the Sky Bird Appearance package in profile, but strangely depicted a Sky Bird with Rally II wheels, instead of standard cast-aluminum type. Also shown were the non-Trans Am dashboard with the new three-spoke steering wheel and tilt-wheel option, blue velour bucket seat, and AM/FM radio and separate eight-track tape deck unit.

to enhance its appearance. The package came in two different variations: Code Y81 with a conventional roof was $556 extra, and code Y82 with a Hurst Hatch (T-top) roof treatment was a $1,143 option. Despite the rather steep price of the T-top, it was added to nearly 14,000 cars, while the non-T-top Special Edition had a production run of under 2,000 units. If nothing else, this was a clear sign of the Trans Am's growing upscale appeal.

All Firebirds continued to be VIN coded with the same system used since 1972. There were 13 characters stamped in the plate located in the left-hand, upper surface of the instrument panel. The sixth was changed to a 7 for 1977. The fifth character varied, since it identified the type of engine the car left the factory with. These symbols adhered to the following scheme:

Engine	Displacement	Carb.	Exhaust	Code	Source
V-6	231 ci	2V	single	C	Buick
V-8	301 ci	2V	single	Y	Pontiac
	350 ci	4V	single	L, R	Pontiac
				X	Oldsmobile
	400 ci	4V	dual	Z	Pontiac
	403 ci	4V	single	K	Oldsmobile

Displacement	Horsepower	Transmission	Code
350 ci	170	manual	—
		Hydra-matic	YA, YB, Y9, XB, XC
400 ci	180	manual	—
		Hydra-matic	XA, XD, XF, XH, XJ, XK, Y4, Y7, YG, YU
	200	manual	WA
		Hydra-matic	Y6

Sales and production of all Firebird models continued to increase in 1977. The largest gains were in the upscale, luxury segment of the market where the Esprit and Trans Am had great appeal. Even the base Firebird saw a healthy improvement, but the Formula—which was geared strictly to the performance enthusiast—realized only a modest increase. Total Firebird production topped 150,000 for the first time:

A completely new hood, retaining the twin-scoop Formula trademark, was designed. That year, a rear decklid spoiler became standard equipment on this Firebird model. The new, small hubcaps (also used on base models) were standard, but this Formula shows off the snowflake cast-aluminum rims.

Name	VIN	Engine	Code	Manual	Automatic	Total
Base Firebird	S87	—	—	—	—	30,642
Esprit	T87	—		—	—	34,548
Formula	U87	—		—	—	21,801
Trans Am	W87	400 ci	L78	—	29,313	29,313
			W72	8,319	10,466	18,785
		403 ci	L80	—	5,079	5,079
Trans Am						
Special Edition (Y81)	W87	400 ci	L78	—	748	748
			W72	384	549	933
			L80	—	180	180
Trans Am						
Special Edition (Y82)*	W87	400 ci	L78	—	6,030	6,030
			W72	2,699	3,760	6,459
		403 ci	L80	—	1,217	1,217
Total						155,735**

*Includes T-top

**15,080 cars were produced w/V-6 powerplants. There is no breakout available between Firebird V-6s and Esprit V-6s.

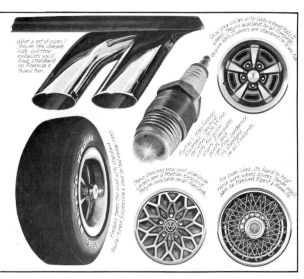

The 1977 Firebird sales catalog illustrated that side splitter exhausts were standard on Formulas and Trans Ams. All Firebirds had electronic ignition. Rally II rims (upper right) were available on all models; standard on Trans Am. Wire wheel covers (lower right) were available for all but Trans Am. Cast-aluminum wheels (bottom center) were available on all Firebirds with silver or body-color accents. White-letter, steel-belted radial tires were used on Trans Ams, and could be ordered for other models.

Just a glance at the above figures tells you that some of the 1977 packages had extremely low production rates. For example, only 180 Y81 Special Edition models were produced with the 403-ci engine. Potential buyers should keep in mind that such rarity has to be measured against the appeal of the W72 performance-engine option. For example, it's likely that one of the 384 Y81 Special Editions made with the W72 engine and four-speed manual transmission would be more desirable to the knowledgeable Firebird collector than the same car with the L80 engine and Turbo Hydra-matic—even though the latter combination is more rare. In addition, a Y82 Special Edition with the T-top might be even more desirable, although its production was substantially higher.

To be a wise buyer, you have to be able to tell what the various option codes mean, as well as how many cars were made with the option. The above chart should be only the starting point for forming your buying decisions.

People who purchased and drove 1977 Firebirds noticed some other "typical" problems with the cars. In addition to the premature rusting, poor paint quality, and loose body panels that had afflicted earlier models, these cars were also frequently troubled by electrical system problems and difficulties with manual transmissions.

The gas mileage figures recorded for the 1977s weren't greatly improved, either. The V-6 usually gave 17.5 to 18.5 miles per gallon, or one mile per gallon more than the inline six. The 301-ci V-8 was good for 15.5 to 16.5 miles per gallon. With four-barrel carburetors, figures for the bigger-cube engines were naturally lower. With the 350-ci engine, 14 to 15 miles per gallon was average; with the 400-ci engine, 13.5 to 15 miles per gallon; and with the 403-ci engine, 14 miles per gallon. However, the increase in sales suggests that the buyers did not really care about economy ratings.

Motor Trend magazine road tested the 1977 Firebird Formula and said, "Detroit's putting some of the fun back." The car had the 301-ci engine (the magazine said it was "closer to 302 [cubic inches]") and four-speed manual transmission. According to the report, it offered "the amalgamation of uncommonly good handling with a frugal new V-8."

The specification chart put fuel economy for *Motor Trend's* 73-mile "editorial" test loop at a surprisingly high 20.9 miles per gallon. Performance was given at 12 seconds for 0–60 miles per hour and 17.9 seconds for the quarter mile, with a terminal speed of 75.2 miles per hour.

The report concluded that the car ($6,900 with a good assortment of options) was good looking, well built, and fun to drive. "This car rates the maximum number of stars when it comes to handling and general road manners," added *Motor Trend's* technical editor, John Ethridge.

There was one factory recall for 1977 Firebirds. It was made to check that the locking tabs on the steering shaft hadn't been bent over or riveted improperly. If the bend was incorrect or the rivets were not properly peened, there was a chance of sudden failure with loss of steering control. This is an important thing to check for and correct, in case the proper repairs weren't made years ago.

The redesigned front end for 1977 Firebirds (left) featured a pointed nose, new Endura bumper panel, and dual headlights with rectangular lenses. This look was adopted from the Banshee show car (right). Actually, a series of Firebird show cars used the Banshee name, and this latest edition inspired the 1977 appearance of all Firebirds.

1977 Firebird Options and Accessories

Sales Code and UPC	Description	Retail Price ($)
LD	231-ci V-6 2V (S87, T87)	std.
L27	301-ci V-8 2V (req. w/M20, M40; S87, T87)	65.00
L34	350-ci V-8 4V	
	S87, T87	155.00
	U87	90.00
L76	350-ci V-8 4V	
	Calif. only; S87, T87	155.00
	Calif. only; U87	90.00
L78	400-ci V-8 4V (req. w/M40 trans; U87)	155.00
W72	400-ci V-8 4V	
	Req. w/M20, M40 trans.; U87	205.00
	Req. w/M20, M40 trans.; W87	50.00
L80	403-ci V-8 4V (U87)	155.00
M20	4-speed man. trans. (L27/W72 only; S87, T87)	257.00
M40	Turbo Hydra-matic auto. trans. (S87, T87)	282.00
C60	Air conditioning	478.00
W50	Formula Appearance pkg. (U87) only	127.00
W60	Sky Bird Appearance pkg.	
	T87 w/Lombardy trim	342.00
	T87 w/doeskin trim	315.00
Y82	W87 Special Edition pkg. w/T-top	1,143.00
Y81	W87 Special Edition pkg. w/o T-top	556.00
G80	Safe-T-Track differential (std. w/W87)	54.00
UA1	H-D battery (maintenance-free)	31.00

Sales Code and UPC	Description	Retail Price ($)
AK1	Custom belts (std. in T87 w/W60)	21.00
JL2	Power brakes (req. w/all V-8s; S87, T87)	61.00
U35	Elec. clock (n/c w/rally gauges)	21.00
D55	Front console (req. w/M40 trans.; S87, T87)	75.00
D58	Rear console	46.00
K30	Cruise control	80.00
A90	Remote-control deck lid release	18.00
C49	Elec. rear window defroster	82.00
AU3	Power door locks	68.00
VJ9	Calif. emissions equipment	70.00
NA6	High-altitude performance option	22.00
K05	Engine block heater	13.00
Y96	Firm Ride pkg. (n/c w/V81)	11.00
UR1	Economy & vacuum gauges (together w/W63 only; exc. W87)	27.00
W63	Rally gauges & clock (req. w/UR1 w/LD engine; exc. W87)	60.00
WW8	Rally gauges, clock & tach. (w/V8 only; exc. W87)	116.00
A01	Tinted glass (all windows)	50.00
QBX	Rally RTS handling pkg. (req. w/N65; U87 only)	70.00
QCY	Rally RTS pkg.	
	U87	116.00

The 455-ci V-8 was not available in 1977, therefore all Trans Ams came with the 6.6-liter engine. In California, strict emissions regulations forced GM to offer only the Oldsmobile-built 403-ci V-8 (185 horsepower) in the Trans Am. Outside California, two Pontiac-built 400-ci V-8s were available. Standard equipment was the L78 version with 7.6:1 compression and 180 horsepower. The performance option was the W72 with 8.0:1 compression and 200 horsepower.

Sales Code and UPC	Description	Retail Price ($)
	W87	46.00
WW7	Trans Am hood decal (W87 only)	62.00
BS1	Additional acoustical insulation (std. w/T87)	27.00
C95	Dome reading lamp	16.00
Y92	Lamp group package	16.00
B37	Color-keyed front and rear floor mats	18.00
D35	Outside dual sport mirrors (LH remote control)	31.00
D34	Visor vanity mirror	4.00
D64	Illuminated visor vanity mirror	32.00
B84	Vinyl bodyside moldings	40.00
B93	Door edge guard moldings	9.00
B80	Roof drip moldings	
	Std. w/T87; S87, U87 w/o CB7	17.00
	W87	17.00
B96	Wheel opening moldings	
	S87	18.00
	U87 w/o W50	18.00
B85	Window sill & rear hood moldings (std. w/T87)	24.00
V02	Super cooling radiator	
	w/o C60, V81	53.00
	w/C60; w/o V81	29.00

Sales Code and UPC	Description	Retail Price ($)
	w/V81	n/c
U63	AM radio	79.00
U69	AM/FM radio	137.00
U58	AM/FM stereo	233.00
UN8	Citizen's band (CB) radio, 23-channel (req. w/D55, UN9)	195.00
U57	8-track stereo player (req. w/D55, opt. radio)	134.00
UN9	Radio accommodation pkg. (n/c w/opt. radio)	23.00
U80	Rear speaker (w/AM, AM/FM radio only)	23.00
D80	Rear spoiler (S87, T87; std. w/U87, W87)	51.00
NK3	Formula steering wheel	
	S87	61.00
	T87, U87	43.00
N30	Luxury cushion steering wheel (S87)	18.00
N33	Tilt steering wheel	57.00
D98	Vinyl accent strips (all exc. W87)	49.00
CC1	Glass sunroof (dual hatches; w/o Y82)	587.00
CB7	Canopy top (na w/W87)	105.00
V81	Light trailer group	
	w/o C60	64.00
	w/C60	40.00

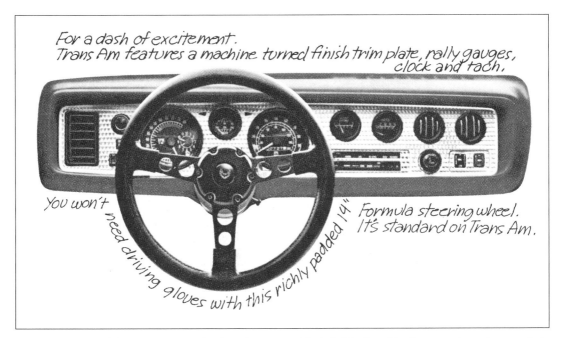

For a dash of excitement. Trans Am features a machine turned finish trim plate, rally gauges, clock and tach.

You won't need driving gloves with this richly padded 14"

Formula steering wheel. It's standard on Trans Am.

An engine-turned instrument-panel trim plate, Rally gauges with tachometer and clock, and drilled-spoke Formula steering wheel were again featured in the Trans Am. Formula wheels were black-finished and had a gear-shaped center hub with small Firebird medallion. This wheel was optional in all other models. However, the Formula wheel was never standard equipment in Formulas.

Sales Code and UPC	Description	Retail Price ($)
Y90	Custom trim	
	Vinyl doeskin; U87, W87	91.00
	Lombardy cloth; U87, W87	118.00
	Regular vinyl; T87 w/o W60	27.00
JL1	Pedal trim packages	
	S87	6.00
	U87, W87; w/o Y90	6.00
P01	Deluxe wheel covers (n/c T87, S87, U87)	34.00
N95	Wire wheel covers	
	na w/W87; S87, U87	134.00
	T87	100.00
YJ8	Cast aluminum snowflake wheels	
	S87	227.00
	T87 w/o W60	193.00
	U87	227.00
	W87	121.00
N98	Argent silver Rally II wheel w/trim rings	
	S87, U87	106.00

Sales Code and UPC	Description	Retail Price ($)
	T87	72.00
N67	Body-color Rally II wheels w/trim rings	
	S87, U87	106.00
	T87	72.00
A31	Power windows (req. w/D55 console)	108.00
N65	Stowaway spare tire	n/c
OKM	FR78-15 blackwall fiber-belted radial	
	S87, T87 w/o N65	(45.00)
	T87, S87 w/N65	(36.00)
OKN	FR 78-15 whitewall fiber-belted radials	
	S87, T87 w/o N65	(4.00)
	S87, T87 w/N65	(3.00)
OBW	FR78-15 whitewall steel-belted radials	
	Exc. W87; w/o N65	41.00
	Exc. W87; w/N65	33.00
OBP	FR8-15 white-letter steel-belted radials	
	Exc. W87; w/o N65	55.00
	Exc. W87; w/N65	44.00

Two versions of the black Special Edition Trans Am were available in 1977. The Y81 had the solid roof. Shown here is the same model, with the Y82 Hurst Hatch, or T-top roof, with removal panels. This feature added nearly $1,200 to the list price and was installed on 14,000 cars. Fewer than 2,000 Special Editions were made with solid coupe roofs. Custom gold decals and gold snowflake wheels really set off this well-integrated appearance option.

Chapter 12

1978

★	**Firebird V-6, V-8**
★	**Esprit V-6, V-8**
★★	**Esprit Sky Bird**
★★	**Esprit Red Bird**
★	**Formula**
★★	**Formula with W50**
★★	**Trans Am**
★★★	**Trans Am**
	Special Edition with Y82
★★★★	**Trans Am**
	Special Edition with Y84
★★★★	**Trans Am**
	Special Edition with Y88

For 1978, the Firebird story was mainly one of refinements, reshuffled engines, redesigned seats, and new, midyear options. A pair of midyear appearance packages are of particular interest to collectors.

While base Firebirds sold in high-altitude counties came with a mandatory 350-ci V-8, the 3.6-liter V-6 as carried over as the base power-plant for other areas. Equipment and appearance changes were virtually nonexistent, except for slight variations in grille styling and switching concealed wipers to the list of standard features.

The standard interior had new door trim panels with smaller armrests and vertical ribs limited to the center section. Color changes were made in upholstery materials, too. Base price climbed to $4,545.

Specifications for the V-6 were unchanged. A new, 305-ci (5.0-liter) V-8 replaced the 301-ci option, but was not offered in California. The new Chevrolet engine had a two-barrel carburetor and an 8.4:1 compression ratio. With a four-speed manual transmission attached, it was rated at 145 net horsepower at 3,800 rpm and 245 foot-pounds at 2,400 rpm. With the Turbo Hydra-matic, corresponding figures were 135 net horsepower at 3,800 rpm and 240 foot-pounds at 2,000 rpm.

An all-new, 350-ci (5.7-liter) V-8 was also available for base Firebirds. Also built by Chevrolet, it had a bore and stroke (4.00x3.48 inches) that was different than either 350 used in 1977. The carburetor was a four-barrel and the compression ratio was 8.2:1. With a four speed, it was rated at 170 net horsepower at 3,800 rpm and

270 foot-pounds at 2,400 rpm. With an automatic, comparable figures were 160 net horsepower at 3,800 rpm and 260 foot-pounds at 2,400 rpm.

With prices starting at $4,842, the Esprit's only real update for 1978 was a new Custom interior. Seat upholstery featured new rear shells. The bolsters had plain panels with horizontal ribs at the top and bottom of the seatbacks and the rear of the cushions. Door trim panels no longer carried embossing in the center nor pull straps. Some color names were changed, as well.

Esprit power teams matched those for the base Firebird. Transmissions, for cars in both lines, included the three-speed that was standard with V-6s, or the four-speed manual or Turbo Hydra-matic with specific V-8s.

The Sky Bird Appearance package continued to be offered early in the season. At midyear, it was replaced by a new Red Bird package that included two-tone red paint, carmine red Custom interiors (vinyl or cloth), a red Formula steering wheel, red grille liners and taillight bezels, and red-trimmed snowflake wheels. There were also special Red Bird decals for the sail panels, but these were gold, as were the lower-body perimeter pinstripes.

Like the blue Sky Bird package, the Red Bird option was attractive and special looking. Unfortunately, the number of cars built this way isn't recorded, but the neat, integrated look of the cars adds some degree of collector appeal.

The 1978 Formula had a new type of black-accented grille and came standard with Rally II wheels and trim rings (replacing standard, small hubcaps). The new interior changes were used for Formulas, too. Custom interiors were still optional in the line.

Base price for the 1978 Formula was $5,448, which included the 305-ci two-barrel V-8, but the 350-ci four-barrel engine was required (at additional cost) in high-altitude areas. A 400-ci four-barrel powerplant (requiring the Turbo Hydra-matic transmission) was optional for federal certification areas only. With four-speed attachments, this engine utilized 8.1:1 compression and gave 220 net horsepower at 4,000 rpm with 320 foot-pounds of torque at 2,800 rpm. With Turbo Hydra-matic, the specs were 7.71:1 compression, 180 net horsepower at 3,600 rpm and 325 foot-pounds of torque at 1,600 rpm.

In California, the top Formula option was the Olds-built 403-ci engine. It had a slightly lowered 7.9:1 compression ratio and developed the same maximum torque at 2,000 rpm as in 1977.

Top dog in the Firebird line-up was still the Trans Am. About the only changes in it were the few interior trim revisions and the fact that the Rally gauge cluster with tachometer and clock was no longer standard. Instead, the tach-only cluster was standard and the tach and clock cluster was turned into a no-charge option.

Trans Am had exterior color selections expanded to eight possibilities instead of six. The black-with-gold trim Special Edition came only with a T-top. There were two versions of this package: one was the Y82 built at Norwood, Ohio, and the other was the Y84 built at Van Nuys, California.

The California-built examples are rarer. More than 10 times as many were made in Ohio. The Norwood plant installed only 400-ci V-8s, while the California factory built 210 of these cars with the 403-ci powerplant. There were also two types of T-tops used: the original Hurst Hatch and a new, improved Fisher version.

The Fisher T-tops were larger and heavier, extending more to the rear and toward the center. They had black-finished perimeter trim while most, but not all, Hurst Hatches had bright trim. According to owners, the Hurst Hatch tended to leak and it was eventually phased out of production.

A new Special Edition package, released at midyear, came only with the Fisher T-top. This was the Y88, or Gold Special Edition. This option made its debut at the Los Angeles Auto Show in January. The body was done in solar gold, as were the snowflake wheels. A camel tan interior was also included, along with a new, dark gold "big bird" hood decal and striping.

Sales of the Gold Special Edition were very strong, but the paint used on California-built units proved to be a problem. Because of the state's stricter environmental laws, a water-based paint was used. It was not as lustrous or glossy as the standard paint used at Norwood, Ohio. Many cars came off the California assembly line with paint flaws or bodies that did not match the color of the Endura rubber end panels. Paint touch up became a major problem.

The fundamental Firebird. Great value on any street.

For 1978, the base Firebird coupe body-color bumpers were continued front and rear. The Firebird crest appeared on the roof sail panels and the center of the decklid. The Pontiac name was on the right-hand corner of the decklid, and Firebird lettering was placed on the front fenders.

Dealers had to repaint many cars under warranty. Today, a nicely refinished or restored example might have extra appeal to a collector, but an original Gold Special Edition made at Van Nuys has to be rated a lesser investment than the Black Special Edition.

Engine selections for 1978 Trans Ams were basically the same as in 1977, except for a few minor changes. The L78 powerplant received a very slight compression increase to 7.7:1. The W72 option gained 20 net horsepower and lost 5 foot-pounds of torque because of a new 8.1:1 compression ratio (up from 8.0:1). Compression for the L80 engine dropped slightly (to 7.9:1) and torque was still 320 foot-pounds, but at 2,000 rpm.

Base price for the Trans Am was up to a hefty $5,799. The L80 engine was a mandatory installation in California-built cars at no extra charge. The base L78 and high-performance W72 engines were not available in the California market. Only the latter powerplant was still offered with four-speed manual transmission attachments.

There was a slight change in GM's VIN coding system this year. The first character was still a 2 for Pontiac. Next came the model number: S87 for base Firebird, T87 for Esprit, U87 for Formula, and W87 for Trans Am. The fifth character was either a letter or number indicating the type of engine (Firebirds continued to use only letter codes for engines, however). The sixth character was an 8 for 1978. The seventh was a letter indicating the assembly plant. The last six digits were the sequential production number, running upward from 100001.

The fifth character (engine codes) appearing in Firebird VINs was as follows:

Code	Engine	Displacement	Source
A	V-6	231 ci	Buick
U	V-8	305 ci	Chevrolet
L, R	V-8	350 ci	Chevrolet
X	V-8	350 ci	Chevrolet
Z	V-8	400 ci	Pontiac
K	V-8	403 ci	Oldsmobile

Engine block codes were found in the normal location on Pontiac built 400-ci engines. The code identified both engine and transmission attachment at follows:

Engine	Displacement	Horsepower	Transmission	Code
V-8	400 ci	180 nhp	manual	—
			automatic	XJ, XK, YK, YH, YS, YT, X9, YJ, YR, YW, YA,YU,
		200 nhp	manual	WC
			automatic	X7

The standard vinyl bucket-seat interior had new door panels for the first time since 1970 1/2. This car has the automatic transmission console without power window switches, since the standard windows were conventional, hand-crank type. Note the smaller door armrests.

Custom interiors were revised completely in 1978. There were fewer pleats, and they ran across the seats at two places on the backrests and at one place on the cushion. Also changed drastically was the patterning of the interior door panels. The Custom interior was still standard on all Esprits and done in light-blue monotone in the Sky Bird and a red monotone on the new Red Bird.

The 1978 Firebirds had about the same overall repair rate as 1977 models. Typical problems were about the same, with the gold paint used on the Y88 Special Edition being another potential trouble spot. Cars featuring Hurst T-tops should always be carefully inspected for signs of sealing problems with the removable roof hatches, as severe body and floor rust may result from leaking hatch panels. It would be a wise move to carefully inspect the underbody of any car so-equipped prior to purchase; rear footwells are especially affected.

Fuel economy rating for carryover engines were also similar to those for comparable 1977 power teams. The new, Chevy-built 305-ci motor was found to be even more frugal than the 301-ci V-8 that it replaced. It was rated at a 16 miles per gallon average. The 350-ci V-8 from Chevrolet was rated at 15 miles per gallon.

There was one factory recall for 1978, linked to the possibility that the fan blade hub could suffer fatigue. When this occurred, the fan blade spider could break apart, allowing a two-blade segment to be thrown off.

A November 1977 article in *Car and Driver* described the Trans Am 400 as "very sophisticated and impeccably well mannered." The magazine's test car used the 200 net horsepower engine with Turbo Hydra-matic. Acceleration was charted at 0–110 miles per hour in 34.8 seconds.

In the May 1978 issue of *Car and Driver,* a Trans Am with the same engine and a four-speed was evaluated by Assistant Editor Michael Jordon. It was his choice for the Year's Best Car. Jordon described it as "hokey" looking because of its external "street racing clichés," but added:

Velour Custom trim options were available on all Firebirds at extra cost. Here is the interior in gold on a Trans Am with the black Formula wheel and engine-turned instrument-panel trim. Note that the console is designed to accept the four-speed manual transmission shifter and also incorporates power window controls.

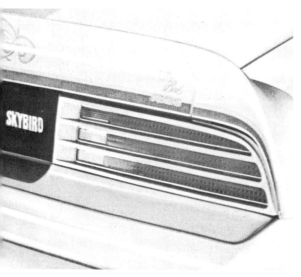

For both Sky Birds and Red Birds, a flowery decal appeared at the center and right-hand corner of the rear spoiler. The decal above the triple-deck taillight lens carried the name of the special package.

Available at extra cost on all 1978 Firebirds was a digital quartz clock and AM/FM radio with electronic readouts for station frequency, time, elapsed time, and date. This was part of one of the nine different sound systems offered, and it was manufactured by GM's Delco Division.

"beneath this lies a core of street-racing hardware: a 220 [horsepower] engine, limited-slip differential, big tires, and a four-speed transmission that will give you tennis elbow in just one dust-up on your favorite bit of mountain road."

The Firebird continued to be increasingly popular during 1978, with the sales of all four car lines spiraling upward. The popularity of the V-6 declined slightly, with a total of 13,595 base Firebirds and Esprits using this powerplant. There is no breakout available for V-6 installations by car line. Due to the changes in availability of optional engines and transmissions, some combinations saw very limited production. For instance, at the Van Nuys, California, plant only 20 cars were built with the Y84 (Black Special Edition) package, 400-ci L78 engine, and four-speed manual transmission.

The total known production picture for 1978 models looks like this:

Name	VIN	Engine	Code	Manual	Automatic	Total
Base Firebird	S87	—		—	—	32,671
Esprit	T87	—		—	—	36,926
Formula	U87	—		—	—	24,346
Trans Am Coupe	W87	400 ci	L78	6,777	57,035	63,812
			W72	4,112	4,139	8,251
		403 ci	L80	—	8,969	8,969
Trans Am Black Special Edition (Y82)		400 ci	L78	489	2,856	3,345
Trans Am Black Special Edition (Y84)			L78	20	68	88
(Y84)		403 ci	L80	—	210	210
Trans Am Gold Special Edition (Y88)		400 ci	L78	1,267	6,519	7,786
(Y88)		403 ci	L80	—	880	880
Total						187,284

The 1978 Esprit with Red Bird Appearance package was essentially made up of the same ingredients as the Sky Bird package, but with two-tone red exterior (dark red below perimeter line), red velour interior, and special gold decals on the sail panels and rear. Cast-aluminum snowflake wheels were accented to match each package, and the grille had colored inserts.

1978 Firebird Options and Accessories

Sales Code and UPC	Description	Retail Price ($)
LD5	3.8-liter V-6 2V (S87, T87; n.a. w/NA6)	std.
	5.7-liter V-8 4V	
LG3	5.0-liter V-8 2V	
	S87, T87	117.00
	U87	n/c
LM1	5.7-liter V-8 4V	
	S87, T87	265.00
	U87	115.00
L78	6.6-liter V-8 4V	
	(na w/NA6, VJ9; req. w/auto. trans.)	
	Formula (U87)	205.00
	Trans Am (W87)	n/c
W72	6.6-liter V-8 4V	
	(na w/NA6, NJ9; req. w/auto trans. or M214-speed manual)	
	Formula (U87)	280.00
	Trans Am (W87)	75.00
L80	6.6-liter (403 ci) V-84V (w/NA6, VJ9 only; req. w/automatic trans.)	

Sales Code and UPC	Description	Retail Price ($)
	Formula (U87)	205.00
	Trans Am (W87)	n/c
MM3	3-speed man. trans. w/floor shift (w/LD5 only)	n/c
MM4	4-speed man. trans. (w/LG3; LM1; n.a. w/NA6, VJ9)	
	Formula (U87)	(182.00)
	S87, T87	125.00
M21	close-ratio 4-speed man. trans. (w/W72 only; n.a. w/NA6, VJ9)	
	Formula (U87)	n/c
	Trans Am (W87) only	n/c
MX1	3-speed auto. trans. (req. w/D55)	
	S87, T87	307.00
	U87, W87	n/c
C60	Custom air conditioning	508.00
K76	H-D alternator 61-amp. (avail. only w/LG3, LM1)	
	w/o C49, C60	31.00

A new type of black-accented grille was one of the Formula Firebird's identification features in 1977. Available engines included 305-, 350-, and 400-ci blocks, but the engine displacement was no longer used as part of the Formula model name.

Sales Code and UPC	Description	Retail Price ($)
	w/C49, C60	n/c
K81	H-D alternator 63-amp. (avail. only w/LD5, L78, W72, L80)	
	w/o C49, C60	31.00
	w/C49, C60	n/c
W50	Formula Appearance pkg. (U87 only)	137.00
W60	Sky Bird Appearance pkg.	
	T87 w/velour Lombardy cloth	461.00
	T87 w/vinyl doeskin trim	426.00
W68	Red Bird Appearance pkg.	
	T87 w/velour Lombardy cloth	465.00
	T87 w/vinyl doeskin trim	430.00
Y82	Black Special Edition (W87 only; built at Norwood)	1,259.00
Y84	Black Special Edition (W87 only; built at Van Nuys)	1,259.00
Y88	Gold Special Edition (W87 only)	1,263.00
G80	Safe-T-Track rear axle (S87, T87, U87)	60.00
UA1	H-D battery	20.00
AK1	Custom belts	
	All exc. Sky Bird, Red Bird	20.00
	Sky Bird, Red Bird	n/c
U35	Electric clock (na w/Rally gauges)	22.00
D55	Console	
	Req. w/auto. trans.; S87, T87	80.00

Sales Code and UPC	Description	Retail Price ($)
	U87, W87 w/auto. trans.	n/c
K30	Cruise control (req. w/auto. trans.; w/LG3, LM1, L78, W72)	90.00
A90	Remote-control deck lid release	21.00
C49	Electric rear window defroster	92.00
VJ9	California emissions systems	75.00
NA6	High-altitude emissions system	33.00
K05	Engine block heater	14.00
W63	Rally cluster & clock (exc. W87)	63.00
WW8	Rally cluster w/clock & tach.	
	Exc. W87	123.00
	W87	n/c
A01	Tinted glass, all windows	56.00
CC1	Fisher hatch roof	
	w/o Y84, Y88	625.00
	w/Y84, Y88	n/c
WY9	Hurst Hatch roof	
	w/o Y82	625.00
	w/Y82	n/c
WW7	Hood decal (W87 w/o Y82, Y84, Y88. nc w/Y82, 84, 88)	66.00
BS1	Added acoustical insulation (nc w/T87)	29.00
C95	Dome reading lamps	18.00
Y92	Lamp group	18.00

The Trans Am sported the slightly revised 1978-design grille. Inside, a clock was no longer standard as part of the Rally gauge cluster, although a tachometer was. Buyers of Trans Ams (and also Formulas) could substitute snowflake wheels for the standard Rally IIs at no additional cost.

Sales Code and UPC	Description	Retail Price ($)
B37	Front & rear mats; color-keyed	21.00
D35	Mirror group (S87)	34.00
D34	Visor vanity mirror	5.00
B84	Roof drip moldings (na w/WY9, CC1; w/o CB7)	
	w/S87, U87	20.00
	w/W87	20.00
	w/T87	n/c
B96	Wheel opening moldings	
	na w/W87; w/S87, U87; w/o W50	21.00
	U87 w/W50; T87	n/c
B85	Windowsill and rear hood moldings (std. w/T87)	25.00
JL2	Power brakes (req. w/V-8 & V-6 w/C60)	
	w/S87, T87	69.00
	w/U87, W87	n/c

Sales Code and UPC	Description	Retail Price ($)
AU3	Power door locks	80.00
A31	Power windows (req. w/D55)	124.00
V02	Super cooling radiator	
	w/C60	31.00
	w/o C60	56.00
U63	AM radio	83.00
UM1	AM radio w/8-track player	233.00
U69	AM/FM radio	154.00
UP5	AM/FM radio w/40-band CB (req. w/U80)	436.00
U58	AM/FM stereo radio	236.00
UN3	AM/FM radio w/cassette	351.00
UY8	AM/FM stereo w/digital clock (na w/W87, w/U35)	392.00
UP6	AM/FM stereo radio w/40 channel CB	518.00
UM2	AM/FM stereo radio w/8-track player	341.00
UN9	Radio accommodation pkg. (nc w/opt. radios)	27.00

This 1978 Trans Am has several uncommon add-ons. They include the mid-bodyside molding package and optional stripes along the tops of the doors and front fenders. (The wheels and tires shown here are aftermarket items.) In 1978, the hatch roof came from two sources: Hurst Performance Company and GM's Fisher Body Division. There are slight differences in the two types.

Sales Code and UPC	Description	Retail Price ($)
U80	Rear seat speaker (w/UP5, U63, U69 only)	24.00
D80	Rear deck spoiler	
	S87, T87	55.00
	U87, W87	n/c
NK3	Formula steering wheel	
	S87	65.00
	T87 w/o W60	46.00
	T87 w/W60, W87	n/c
	U87	46.00
N30	Luxury steering wheel (S87; nc w/T87, U87; n.a. w/W87)	n/c
N33	Tilt steering	72.00
D98	Vinyl accent stripes (na w/W87, W50, W60)	52.00
C87	Canopy top (na w/W87; incl. B80)	111.00
WS6	Trans Am special performance pkg.	
	w/W72	324.00
	w/L80	249.00
Y90	Custom trim group	
	w/doeskin vinyl (U87, W87)	99.00
	w/Lombardy cloth (U87, W87)	134.00

Sales Code and UPC	Description	Retail Price ($)
	w/Lombardy cloth (T87 w/o W60)	35.00
	w/Lombardy cloth (T87 w/W60)	n/c
P01	Deluxe wheel covers (S87)	38.00
N95	Wire wheel covers	
	S87	146.00
	T87	108.00
YJ8	Painted cast aluminum wheels, four	
	w/S87	290.00
	T87 w/o W60	252.00
N98	Argent silver Rally II wheels (4) w/trim rings	
	S87 w/o N65	136.00
	T87 w/N65	79.00
	T87, w/o N65	98.00
	U87, W87	n/c
N67	Body-color Rally II wheels w/four trim rings	
	S87 w/o N65	136.00
	S87 w/N65	117.00
	T87 w/N65	79.00
	T87 w/o N65	98.00
	U87, W87	n/c
CD4	Controlled-cycle windshield wipers	32.00

1978 was the first year for the gold Trans Am Special Edition. These cars enjoyed strong sales and generally high popularity.

Sales Code and UPC	Description	Retail Price ($)
N65	Stowaway spare tire	n/c
QBU	FR78-15 black sidewall steel belted radials (S87, T87)	n/c
QBW	FR78-15 white sidewall steel-belted radials	
	S87, T87 w/N65	37.00
	S87, T87 w/o N65	46.00

Sales Code and UPC	Description	Retail Price ($)
QBP	FR78-15 white-letter steel-belted radials	
	S87, T87 w/N65	49.00
	S87, T87 w/o N65	61.00
QBX	GR70-15 black sidewall steel-belted radials	n/c
QCY	GR70-15 white sidewall steel-belted radials (U87, W87)	51.00

Chapter 13

1979

★	**Firebird V-6, V-8**
★	**Esprit V-6, V-8**
★★	**Esprit Red Bird**
★	**Formula**
★★	**Formula with W50**
★★	**Trans Am**
★★	**Trans Am with T-top**
★★★	**Trans Am Special Edition**
★★★★	**Trans Am Special Edition 10th Anniversary**
★★★★★	**Trans Am Special Edition 10th Anniversary with L78, W72**

The third and last facelift for second-generation Firebirds appeared with the announcement of 1979 models. Pontiac described its new grilleless look as "a broad new forefront cast in durable urethane."

Used on all Firebirds, the treatment featured a more gently contoured, slanted front nose panel with twin, rectangular headlamps set into individual, squarish ports at either end. The center of the panel came to a rounded, V-shaped peak.

Below this was another (bumper) panel with a large license-plate recess in its center and two long, thin, horizontal openings on either side. The openings were filled with six grille lo uvers and the white-lens parking lights. Formulas and Trans Ams featured black-out-style taillamps. Base Firebirds and Esprits had full-width taillights with dark-finished bezels, but they lacked the black, opaque covers used on the other models over the red lenses.

Also new for the year were redesigned deluxe wheel covers and a more plush Custom cloth interior option. Featuring hobnail cloth cushions and seatback inserts, this trim was available for all models except the base Firebird. Initially, special exterior trim packages were limited to the Red Bird option, the Formula Appearance package, and the Black Bird (a black-with-gold-trim Special Edition). The latter was available with or without the Fisher T-top, although production records mistakenly show all cars had the option.

The tenth year for Trans Ams suggested a natural promotional possibility. On February 15, 1979, in conjunction with the Daytona 500 Winston Cup stock car race, a 10th Anniversary option was announced. This package drew 7,500 orders by year's end.

The popular 301-ci V-8 was reissued in two variations for use with specific 1979 models. Other engines were the same as used the previous year, with only minor changes to the horsepower and torque ratings of the 305-ci V-8. Transmission choices were further restricted in California and high-altitude applications.

New options included four-wheel disc brakes for Trans Ams and Formulas. Other changes included column-mounted headlamp dimmers, convex glass in right-hand outside rearview mirrors, and tri-band power antennas with CB radios.

Exterior color choices were cut from 18 to 13, but Trans Ams gained one choice for a total of nine. The new Custom cloth interior option for Esprits, Formulas, and Trans Ams came in black, blue, camel tan, or carmine red.

A base Firebird now listed for $5,076. The standard equipment list was unchanged, but there were several slight dimensional changes. Overall width dropped from 73.4 to 73 inches, wheelbase went from 108.2 to 108.1 inches, and front track increased from 60.9 to 61.3 inches.

The 3.8-liter V-6, carried over without change, was standard in base Firebirds—except in high-altitude areas, where the 5.7-liter V-8 was a required option, at extra cost. Other engines available in base Firebirds were the 4.9-liter V-8 with two- or four-barrel carburetion and the 5.0-liter V-8 with a two-barrel. The drivetrain options that buyers could order depended on where they lived.

With V-6 power, the standard-equipped Esprit had a $5,454 window sticker when introduced in the fall. Appearance distinctions of this model were the same as in the past. The deluxe wheel covers were of the new design. They again had a red circular center medallion with the V-shaped Pontiac insignia. The former triple rings of round openings decreasing in size toward the center were replaced by thin, rectangular openings (angled toward the center) arranged in a single band around the outer perimeter of the covers.

Esprit powertrain options were the same as for base Firebirds at the same prices.

The Red Bird package was shown in the 1979 sales catalog and seemed particularly attractive with the new front styling. Gold, double pinstriping was used liberally to highlight the front

and side edges of the roof, the beltline, cowl line, both sides of the hood center crease, the lower front panel openings, and the entire lower-body perimeter. A two-tone red finish, red-accented snowflake wheels, and a red interior were other features.

Base priced at $6,380, the 1979 Formula had several changes in the list of standard equipment that separated it from other models. No longer included was a rear deck lid spoiler, which now cost extra. Also, a Formula steering wheel was substituted for the luxury cushion type. The Rally gauge cluster with clock was made standard, as was a Trans Am-style engine-turned instrument panel trim plate. Standard tires were switched from GR70-15 blackwalls to 225/70R-15 steel-belted radial blackwalls.

Standard Formula identification features included Firebird decals on the roof sail panels, a bird decal on the rear panel, and the word Formula carried in block letters on a small decal on the lower perimeter feature line, just behind the front wheel openings. The lower-body perimeter was also subtly pinstriped.

The 4.9-liter two-barrel V-8 engine was standard fare in Formulas for federal certification

Pontiac's 1979 sales catalog gave a bird's-eye view of all four Firebirds, with base coupe at the top.

The Red Bird package was continued for 1979, again in two-tone red with gold decal accents. On this car, the V-shaped Pontiac emblem—traditionally rendered in red plastic—was changed to gold. The snowflake wheels had red accents.

areas. The 5.7-liter (high-altitude) and 5.0-liter (California) V-8s were required, at extra cost, in base-equipped cars for certain marketing areas. (Full details appear later in this chapter.) The 400- or 403-ci V-8s were options in specific areas. The three-speed manual transmission was not available anywhere.

Offered once again was the W50 Formula Appearance package, which was promoted in the sales catalog as an option designed "to bring the brawn to the surface." It included contrasting lower-body perimeter finish with the breakline accented with a broad stripe of a third color, which was repeated along the bottom of the tail-lamp panel. With this option, a rear deck lid spoiler was standard equipment.

In the sales catalog, the W50 treatment was shown on a silver car with a black lower perimeter. The stripes were red. The word "Formula" appeared above the taillights in bold red letters with silver core accents, and on the lower-body perimeter in silver letters with black core accents. Also included were three-tone vinyl accent stripes for the upper edge of the simulated hood air scoops.

New for 1979 was a plusher Custom cloth interior option featuring hobnail cloth cushion and back-rest inserts. It could be ordered with all regular models except the base coupe. Of course, integrated appearance packages such as the Red Bird Esprit or 10th Anniversary Trans Am precluded use of this option, too.

The Formula again came with a twin-scoop hood and Rally IIs as standard fare.

The new four-wheel disc brakes had 11-inch rotors. They could be ordered separately or as part of the WS6 Special Performance package available for Formulas and Trans Ams.

Trans Am prices for 1979 began at $6,699. The extras included were the same as in the past. The base powerplant in all areas was an Olds-built 6.6-liter V-8, with the 4.9-liter and 5.6-liter V-8s optional outside of California.

There were changes in other available equipment, most of them quite minor. For instance, white-letter tires could again be had apart from the WS6 package (in 1978, ordering this package was the only way to get white-letter tires).

Pictured in the 1979 sales catalog was a Special Edition Trans Am with camel tan hobnail Custom cloth interior and T-tops. This package came only in black with gold trim. The trimmings included gold bird decals on the sail panels, gold Trans Am lettering behind the front wheel openings, gold lettering (that read Trans Am: Pontiac) on the upper surface of the left-hand bumper panel, gold "Trans Am 6.6" lettering on both sides of the shaker hood scoop with a "big bird" hood emblem with a bronze base decal, and gold accents and feathers.

The 10th Anniversary package included both appearance and performance extras. Features of an exclusive nature included a larger big bird hood decal, bolder-style hubs, a complete complement of convenience options (including cruise control), and special interior with silver leather seats.

These cars came only in silver with the shaker scoop, bumper panel, window sills, rear window surround, and portions of the roof finished in charcoal gray. Red, white, and charcoal pinstripes accented numerous body panels. Behind the front wheel openings were red Trans Am lettering and 10th Anniversary Limited Edition decals.

Most 10th Anniversary Trans Ams had the code L80 (403-ci) four-barrel engine with a Turbo Hydra-matic transmission. Relatively few were produced with the code L78/W72 combination of a Pontiac-built 400-ci V-8 linked to a four-speed manual transmission.

Since the 10th Anniversary Trans Am came only fully equipped, it was marketed as a separate model with a $10,619 sticker price. One example, photographed with Pontiac's new Division Manager, Robert C. Stempel, served as pace car for the 1979 Daytona 500 stock car race. It carried a special Pace Car decal package.

The following list shows standard equipment on the 10th Anniversary Trans Am not specifically mentioned above: power windows; door locks; remote deck lid release; power radio antenna; four-wheel disc brakes; tinted glass; front and rear floor mats; gray Custom seat belts; lamp group; vanity visor mirror; electric rear

A Formula Appearance package was offered again at extra cost. It included three-tone stripes on the simulated hood scoops and Formula graphics on the lower body perimeter in silver with a red core.

New blacked-out-style taillights with opaque covers were used on 1979 Formulas and Trans Ams. Formulas also had model identification on the rear body panel. This car has the Formula Appearance package and no-cost optional snowflake wheels. Another added feature was the heavy-duty Trans Am suspension, standard in Formulas.

A 1979 Trans Am with the new Firebird nose treatment and optional big bird hood decal. Only the black-with-gold trim Special Edition package was marketed in 1979, so this must be a regular model.

window defroster; dome reading lamp; heavy-duty battery; tilt steering; pedal trim package; quartz-halogen headlights; red instrument panel lighting; air conditioning; digital AM/FM signal-seeking stereo radio with 8-track player; silver-tinted hatch roof, and embroidered crest in seats and door panels. The special 15x8-inch Turbo alloy wheels were from Appliance Wheel Company.

Although it was produced in greater numbers than truly scarce Firebirds, like the Trans Am convertible and the 1973-74 SD-455, the special luxury and appearance features of the 10th anniversary Trans Am have made it a desirable model to collectors. Not long after leaving the assembly lines, some of these cars were advertised in enthusiast magazines for as high as $30,000.

These expectations of instantaneous value appreciation may have been partly inspired by the furor over the Corvette Indy Pace Car package of the previous year. (Asking prices for Pace Car Corvettes were as high as $60,000.) And both cars went through similar up-and-down price shifts before leveling off in value.

Today, the 10th Anniversary Trans Ams are probably worth just slightly more than they sold for originally. However, some Pontiac experts insist that the model's future potential as a collector car must be rated in the excellent category. Thus, I have given the standard L80 version a four-star investment rating, and I have rated the rarer L78/W72 version at five stars.

It should be understood that this doesn't mean the 10th Anniversary is worth as much as the 1969 1/2 convertibles or cars with the Super-Duty engines. Instead, it indicates that the value of a 10th Anniversary L78/W72 will appreciate at about the same rate as the other two, rarer models.

For 1979, Pontiac added a new model number code to the four used in 1978. The code X87 was used to identify the 10th Anniversary, or Limited Edition, Trans Am. Otherwise, the basic VIN coding system was unchanged.

The fifth character, the engine code, used in 1979 production was one of the following:

See page 122

There was also a new rear end panel for the 1979 Trans Am. The taillights featured the innovative black lens appearance, which turned red upon brake pedal application. The base Firebird and Esprit also had the new, full-width taillights with black bezels, but lacked the opaque covers used on Formulas and Trans Ams.

The Trans Am price was up to nearly $7,000 for
1979. That year, all cars with a hatch roof had the
GM-Fisher type installed.

The new Firebird front end featured six louvers in
each of the lower bumper-grille openings. This
nose treatment looked especially nice on the Trans
Am and seemed to blend in well with the front
wheel-opening flares. Use of aircraft as background
in factory publicity photos was an ongoing theme.

Code	Type	Carburetor	Displacement	Source
A	V-6	4-bbl	231 ci	Buick
W, Y	V-8	2-bbl	301 ci	Pontiac
G, H	V-8	2-bbl	305 ci	Chevrolet
L, R, X	V-8	4-bbl	350 ci	Chevrolet
Z	V-8	4-bbl	400 ci	Pontiac
K	V-8	4-bbl	403 ci	Oldsmobile

Engine block codes were found in the normal location on the Pontiac-built 400-ci engines. The only code used this year was WH indicating manual transmission attachments. This powerplant was also known as the Trans Am 6.6 engine. It was used only in cars granted federal certification and required the M21 four-speed manual transmission and WS6 special performance package.

According to owner maintenance surveys, manual transmissions in 1979 Firebirds were somewhat of a sore spot, and the cars had electrical system problems. In addition, complaints of premature rusting, loose body-panel alignment, and poor paint quality were common.

A road test of a 1979 Formula with the Trans Am 6.6 setup was printed in the October 1978 issue of *Motor Trend*. Associate Editor Peter Frey seemed to take a generally positive view of the car, but also noted several shortcomings. He attributed most of these to the fact that it had only 200 miles on the odometer and wasn't fully broken in.

Frey found that the car, which was equipped with four-wheel disc brakes, took 140 feet to stop in a panic stop situation from 60 miles per hour. "This figure isn't quite as good as what we were told to expect," he wrote. Frey also noted rear wheel lockup under maximum braking conditions. He said the optional 3.23:1 limited-slip differential made a "ratchety" sound when going around sharp corners at low speed. In another passage, he stated that the Hurst shifter linkage was "hand-bruising" to operate. He suggested a leather shift handle.

Frey advised driving the car as smoothly as possible, noting "it feels so nimble and the power steering is so light and precise, the temptation is to pitch it into a corner, forgetting . . . the kinetic energy involved." The editor also scored the hatch roof's blackglass panels low for "bathing the driver in a rectangular column of heat on a hot day." He suggested silver or gold mirror glass could be used to prevent this problem. Another thing he did not like was that the car had to be shifted into reverse, with the clutch depressed, in order to start it. Frey

Here's the interior of a 1979 Trans Am with four-speed manual transmission, the new hobnail cloth Custom interior option, and tan Formula steering wheel. The Formula also used this wheel and the engine-turned dash insert as standard equipment in 1979.

On the inside, the 10th Anniversary Trans Am featured silver leather bucket seats with vinyl side-facings, silver door panel trim, and exclusive instrument lighting similar to that used in aircraft.

The 10th Anniversary Trans Am was marketed as a separate model, rather than as an appearance package option. The special hatch roof was silver tinted and standard, as were the turbo alloy wheels made by Appliance Wheel Company.

called this "a cumbersome process."

The tester had no complaints about the Formula 400's performance, which he called "strong." The 220 net horsepower engine was good for a 16.1 second quarter-mile at 83.1 miles per hour. Its 0–60-mile-per-hour time was 9.1 seconds.

According to Pontiac engineers, the new Firebird nose (with slope increased from 16 to 22 degrees) was more aerodynamic and would have produced a positive effect on fuel economy, if the EPA had not changed its testing procedures for 1979 models. Under the new system, all engines but the 403-ci powerplant retained the same economy ratings. The new figure for the 403 was 14 miles per gallon.

There were no factory recalls on 1979 Firebirds. As for production totals, there was a hefty increase in Trans Am deliveries and a modest gain in base Firebird output. The Formula saw a very slight increase, while the Esprit's popularity fell. Here are the figures:

Name	VIN	Engine	Code	Manual	Automatic	Total
Base Firebird	S87	—		—	—	38,642
Esprit	T87	—		—	—	30,853
Formula	U87	—		—	—	24,850
Trans Am coupe	W87	400 ci	L78, W72	2,485	—	2,485
		403 ci	L80	—	48,488	48,488
		301 ci	L37	1,590	7,015	8,605
Trans Am w/T-top	W87	400 ci	L78, W72	2,917	—	2,917
		403 ci	L80	—	30,728	30,728
		301 ci	L37	1,530	3,301	4,831
Trans Am Black Special Edition (Y82)		400 ci	L78	489	2,856	3,345
Trans Am Special Edition w/T-top*	W87	400 ci	L78, W72	1,107	—	1,107
		403 ci	L80	—	30,728	30,728
		301 ci	L37	213	360	573
Trans Am 10th Anniversary	X87	400 ci	L78, W72	1,817	—	1,817
		403 ci	L80	—	5,683	5,683
Total						211,453

*PMD production records indicate that all 1979 Special Editions were equipped with a T-top, but there are some such cars in existence today that do not have this feature. It was a delete option and a few buyers apparently preferred getting a credit to getting a T-top.

The 10th Anniversary package—a model-option in technical terms—could be had with either the base 6.6-liter engine, or the higher-performance Trans Am 6.6 with four-speed transmission.

1979 Firebird Options and Accessories

UPC Code	Description	Retail Price ($)
LD5	3.8-liter V-6 (S87, T87; n.a. w/NA6)	n/c
L27	4.9-liter V-8 2V	
	S87, T87; federal only;	
	req. w/auto trans.	270.00
	U87; federal only; req. w/auto. trans.	n/c
L37	4.9-liter V-8 4V	
	S87, T87; federal only	355.00
	U87; federal only	85.00
	W87; federal only	(195.00)
LG3	5.0-liter V-8 2V	
	S87, T87; req. w/VJ9, auto. trans.	270.00
	U87; req. w/VJ9, auto. trans.	n/c
LM1	5.7-liter V-8 4V	
	S87, T87; req. w/NA6, auto. trans.	425.00
	U87; req. w/NA6, auto. trans.	155.00
L78	Trans Am 6.6-liter V-8 4V	
	U87; federal only; req. w/M21, WS6	370.00
	W87, X87; federal only;	
	req. w/M21, WS6	90.00
MM3	3-speed man. trans.	
	(req. w/V-6; n.a. w/VJ9)	n/c

UPC Code	Description	Retail Price ($)
M21	Close-ratio 4-speed man. trans.	
	S87, T87; req. w/L78, L37	325.00
	U87, W87, X87; req. w/L78, L37	n/c
MX1	Turbo Hydra-matic auto. trans.	
	S87, T87; req. w/D55; n.a. w/L78	335.00
	U87, W87, X87; req. w/D55; n.a. w/L78	n/c
C60	Custom air conditioning	529.00
K81	H-D alternator	
	63 amp.; w/o C60, C49	32.00
	63 amp.; w/C60, C49	n/c
W50	Formula Appearance pkg. (U87 only)	92.00
W68	Red Bird Appearance pkg.	
	T87 w/hobnail cloth trim	491.00
	T87 w/doeskin vinyl trim	449.00
Y84	Special Edition Appearance pkg.	
	W87 w/o CC1	674.00
	W87 w/CC1	1,329.00
G80	Limited slip differential	
	(std. w/W87, X87; on S87, T87, U87)	63.00
UA1	H-D battery	21.00
AK1	Custom seat belts	n/c

UPC Code	Description	Retail Price ($)	UPC Code	Description	Retail Price ($)
U35	Elec. clock (std. w/W63; w/o W63)	24.00	B83	Rocker panel moldings (na w/U87; W87; std. w/T87)	18.00
D55	Console (req. w/auto. trans.; std. in U87, W87; S87, T87)	80.00	B80	Roof drip moldings (w/o canopy top; all exc. T87)	24.00
K30	Cruise control (req. w/MX1; n.a. w/L78)	103.00	B96	Wheel opening moldings (S87 only)	22.00
A90	Remote-control rear deck lid release	24.00	B85	Window sill and rear hood moldings (all exc. T87 (Esprit) & W878 (Trans Am)	
C49	Elec. rear window defroster	99.00		w/Y84 Special Edition Appearance pkg.)	26.00
VJ9	Calif. emissions equipment	83.00	JL2	Power brakes	
NA6	High-altitude emissions equipment	35.00		(req. w/V-8 V-6, C60; std. w/U87, W87)	76.00
K05	Engine block heater	15.00	J65	4-wheel disc power brakes	
W63	Rally gauge cluster w/clock (V-8 only; std. w/U87, S87, T87)	67.00		(U87, W87 w/o Special Edition package)	150.00
U17	Rally gauges w/tach. & clock		AU3	Power door locks	86.00
	S87, T87	130.00	A31	Power window (req. w/D55)	132.00
	Std. w/W87, X87; U87	63.00	V02	Super cooling radiator	
A01	Tinted glass	64.00		w/C60; n.a. w/L80 w/WS6 or NA6	59.00
CC1	Hatch roof w/removable panels (W87 w/o Y84)	655.00		w/C60	32.00
D53	Hood bird decal (W87 w/o Y84)	95.00	U75	Power antenna	
BS1	Added acoustical insulation (S87, U87, W87; std. w/T87)	31.00		w/o opt. Radio; n.a. w/UP5 or UP6	66.00
				w/opt. radio	47.00
C95	Dome reading lamp	19.00	U83	Power antenna, AM/FM tri-band	
B37	Color-keyed floor mats (front & rear)	25.00		w/o opt. Radio	87.00
D35	Dual sport OSRV mirrors; LH remote control (S87 only)	43.00		Std. w/UP5, UP6; w/others	68.00
D34	RH visor vanity mirror	6.00	U63	AM radio	86.00
B84	Vinyl body side moldings	43.00	UM1	AM radio w/integral 8-track stereo player	248.00
B93	Door edge guards	13.00	U69	AM/FM radio	163.00

A limited number of 10th Anniversary Trans Ams were tricked out as NASCAR Pace Cars to be used at the Daytona 500-mile stock car race on February 18, 1979. One such car was used at Daytona that year by Pontiac Division Manager Bob Stempel. This is probably the most valuable type of 1979 Trans Am that any collector could own.

UPC Code	Description	Retail Price ($)
UP5	AM/FM w/CNB (incl. U83)	492.00
U58	AM/FM stereo radio	236.00
UN3	AM/FM radio w/stereo cassette	351.00
UY8	AM/FM stereo radio w/digital clock	402.00
UP6	AM/FM stereo radio w/integral 40-channel CNB (incl. U83)	574.00
UM2	AM/FM stereo radio w/integral 8-track player	345.00
UN9	Radio accommodation pkg.	
	w/o U83, opt. radio	29.00
	w/U83; w/o opt. radio	10.00
U80	Rear seat speaker (only w/UP5, U63, U690)	25.00
UP8	Dual rear speakers (w/UP5, U63, U69; std. w/all others)	38.00
WS6	Special Performance pkg.	
	U87, W87 only; req. w/L37, L78, L80	434.00
	W87 w/Y84	250.00
D80	Rear deck spoiler (S87, T87, U87 w/o W50)	57.00
NK3	Formula steering wheel	
	S87	68.00
	T87 w/o W68	48.00
N30	Luxury steering wheel (S87)	20.00
D98	Vinyl accent stripes (S87, T87)	54.00
CB7	Canopy top (na on W87, X87, T87 w/W68)	116.00
B18	Custom trip group	
	S87, W87, w/vinyl doeskin trim	108.00
	U87, W87 w/hobnail velour cloth trim	150.00
	T87 w/W68, hobnail velour cloth trim	42.00
P01	Deluxe wheel cover (S87)	42.00

UPC Code	Description	Retail Price ($)
N95	Wire wheel covers	
	S87	157.00
	T87	115.00
N90	Cast aluminum wheels (S87)	310.00
N90	Cast aluminum wheels	
	T87 w/o W68	268.00
	T87 w/W68	184.00
N98	Rally II rims, argent silver	
	S87 w/N65	126.00
	S87 w/o N65	146.00
	T87 w/N65	84.00
	T87 w/o N65	104.00
N67	Body-color Rally II rims	
	S87 w/N65	126.00
	S87 w/o N65	146.00
	T87 w/N65	84.00
	T87 w/o N65	104.00
CD4	Controlled-cycle windshield wipers	38.00
N65	Stowaway spare tire	n/c
QBU	FR78-15 blackwall steel-belted radial tires (S87, T87)	n/c
QBW	FR78-15 whitewall steel-belted radial tires	
	S87, T87	48.00
	S87, T87 w/N65	39.00
QBP	FR78-15 white-letter steel-belted radial tires	
	S87, T87 w/o N65	64.00
	S87, T87 w/N65	52.00
QGQ	225/70R-15 blackwall steel-belted radial tires (U87, W87, X87)	n/c
QGR	225/70R-15 white-letter steel-belted tires (U87, W87 w/WS6)	53.00

Chapter 14

1980

★	**Firebird V-6, V-8**
★	**Esprit V-6, V-8**
★★	**Esprit Yellow Bird**
★	**Formula**
★★	**Formula with W50**
★★	**Trans Am**
★★	**Trans Am with T-top**
★★	**Trans Am Special Edition**
★★	**Trans Am Special Edition w/T-top**
★★★★★	**Turbo Indy Pace Car**

Engineering refinements and new option packages were the main changes in 1980 Firebirds. There were some trim and color revisions, too. Production fell for the first time in eight years, and the drop was drastic, with the Trans Am suffering most.

General changes for all models included a lighter-weight, dual-exhaust system and new low-friction ball joints. The powerplant line-up was significantly altered and four-speed manual transmission was canceled (it is quite likely that all 1980 Firebirds had automatic transmissions).

Bucket seats used with the standard interior no longer had pleated headrests; they were done in smooth vinyl instead. A new dark blue upholstery color was added. Standard in all models, for the first time, was a front-seat console. Twelve new colors were offered and others were carried over. Trans Am color choices were expanded to a dozen, although not all were available at the same time. In mid-January, nightwatch blue was made available for the Trans Am, but Tahitian yellow was canceled. Interestingly, one new color was Fiero bronze. (The name Fiero was later used for a two-seat Pontiac sports car.)

A new Yellow Bird package replaced the Red Bird option on Esprits. Then, on October 16, 1979, PMD announced that the Trans Am had been selected as the official pace car for the 64th Indianapolis 500. To commemorate this honor, there was a special Turbo Trans Am Indy Pace Car package that was merchandised as a separate model, as the Daytona Pace Car had been in 1979.

Other new options for 1980 included dual front radio speakers, extended-range dual rear speakers, and an electronically tuned (ETR) AM/FM stereo radio with a seek-and-scan feature. After the beginning of the year, a wheel cover locking package, a turbo-boost gauge package, and an audio power booster were added to the list. New turbo-cast aluminum wheels were available for restricted applications, with the restrictions later changed to allow their sale in the California market. Modified options included a quartz electric clock and dual front and rear speaker package. Also, after November 20, 1979, a Firebird roof rack was available, through GM Parts Division, as a dealer installed option.

There were several running changes in planned or actual equipment availability covered by a series of Car Distribution bulletins issued between the summer 1979 production changeover period and the end of the model year. A few of these changes are historically significant, because they show that equipment listed in some factory literature never made it into a production Firebird.

The following is a summary of important information documented by these bulletins:

• There were initial plans to offer a four-speed manual transmission with the 4.9-liter Trans Am V-8. On June 12, 1979, this transmission was put on hold until October. Ten days later it was dropped.

• Plans to market a base 3.8-liter V-6 with California emissions were delayed several times, then put on hold until December. Ultimately, all California cars used a V-8 as their base engine.

• The name of the 4.9-liter engine with

electronic spark control feature was changed to 4.9-liter E/C V-8 on June 22, 1979.

• After being delayed several times, a January hold was placed on 3.8-liter V-6s teamed with the three-speed manual transmission. This was then extended to late in the model year. Although they were listed as an option for federal cars in factory literature, the manual transmission attachments were never made. Industry journals like Wards Automotive Yearbook 1981 show 100 percent use of automatic transmissions in Firebirds for the 1980 model year.

• A limit was placed on production of Firebirds with hatch roofs, restricting this option to 35 percent of total production.

• A Pontiac-built 4.3-liter (265-ci) V-8, which was listed in factory literature, was placed on hold several times and was ultimately dropped from the options list in April 1980.

• The turbo-boost gauge option and audio power booster were both placed on hold until the end of January.

• There was also a hold placed on production of the turbo-charged V-8 at the Van Nuys and Norwood plants. The California factory was given the green light to start production in October 1979, while the delay at the Ohio plant was scheduled to last until November.

Base price for the Firebird six at introduction was $5,604. In the spring, the prices for Firebirds and Esprits were both increased $344. At the same time, Formula and Trans Am prices rose $300. New standard equipment for the base Firebird coupe included the center console and 205/75R-15 blackwall radial tires. Curb weight was reduced by 14.8 pounds.

The Buick-built base 3.8-liter V-6 had slightly more horsepower (115 at 3,800 rpm). A version of this engine was to be made available for California cars, but it never happened. Consequently, a 5.0-liter V-8 was standard in all California-market Firebirds, although not officially until mid-December. Again, it is believed that all cars had automatic transmissions.

Since the 4.3-liter V-8 was canceled, there were essentially two V-8 options for base Firebirds: one for use in federal marketing areas, and one for California. The federal powerplant was the Pontiac-built 4.9-liter V-8, which had its output reduced to 140 net horsepower at 4,000 rpm. The California engine was the Chevy-built 5.0-liter with 150 net horsepower at 3,800 rpm, up slightly in power from 1979.

Window stickers for Esprits, at fall introduction, were $5,967 for the V-6 powered edition. Wide rocker panel moldings were added to the list of features that were standard over base Firebird equipment. The slightly confused power

The 1980 Firebird series had some new engines and a few different appearance packages. One of these was the Yellow Bird option for the Esprit. It had essentially the same features as the Red Bird treatment, but the finish was two-tone yellow.

Here is the Esprit with the Yellow Bird treatment and extra-cost rear decklid spoiler. There was a decal with the model name and pinstriped bird at the center of the rear, above the gas filler door, as well as on the roof sail panels. Full-width taillamp design was continued for all Firebirds in 1980.

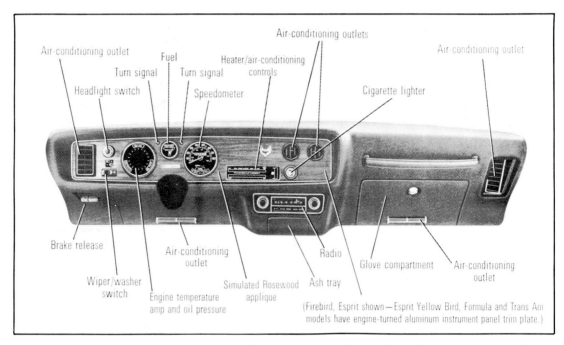

Air-conditioning outlet
Air-conditioning outlets
Air-conditioning outlet
Fuel
Turn signal
Heater/air-conditioning controls
Headlight switch
Turn signal
Speedometer
Cigarette lighter
Brake release
Air-conditioning outlet
Radio
Glove compartment
Air-conditioning outlet
Wiper/washer switch
Engine temperature amp and oil pressure
Simulated Rosewood applique
Ash tray
(Firebird, Esprit shown—Esprit Yellow Bird, Formula and Trans Am models have engine-turned aluminum instrument panel trim plate.)

The Firebird instrument panel and dash was of the same basic design used previously. Circular gauges for temperature and oil pressure, fuel, and speedometer shrunk a little in size. This is the woodgrained style used on base coupes and most Esprits. The Yellow Bird package and Formula and Trans Am models featured the engine-turned aluminum instrument-panel trim plate.

team choices outlined for the base Firebird were also offered for the Esprit.

The Yellow Bird was essentially a Red Bird of a different color with one added touch: black-lens taillights with yellow horizontal ribs were featured. Body color was a special two-tone yellow (code 56) combination. The package included gold striping and Yellow Bird decals for the sail panels and rear spoiler. There was also a gold front end panel crest, yellow snowflake wheels, a black grille, camel tan interior trim, camel seat belts, special emblems and appointments, a gold engine-turned instrument panel appliqué, and originally, a camel tan Formula steering wheel with gold spokes. (Effective November 20, however, steering wheel finish was changed to black-painted spokes on the Yellow Bird and the Trans Am Special Edition.)

Initial prices for 1980 Formulas began at $6,955—more than a 1979 Trans Am! Black taillight lenses and chrome side-splitter tailpipe extensions were added to the previous list of standard Formula features. Literature also specified that the Formula interior included a Rally gauge cluster with quartz electric clock. All else was about the same as in 1979.

The standard powerplant (except in California) was the 4.9-liter V-8 described above. For California, the 5.0 liter (also previously described) was required (this engine came only with air conditioning). When the 4.0 liter was used in Formulas (or Trans Ams), the standard chrome exhaust splitters were replaced with dual resonators and tailpipes.

Two other Pontiac-built 4.9-liter V-8s were new Formula options. They were the E/C version with 155 net horsepower at 4,400 rpm and the turbo, rated at 210 net horsepower at 4,000 rpm.

Teamed with the 155 net horsepower engine was a special-performance automatic transmission featuring a lock-up torque converter. Both engines were limited to cars sold outside California.

The turbocharger used with this engine was made by Air Research Corporation (model TB0305). This unit was also used on some four-cylinder Ford and V-6 Buick powerplants in 1980. It was sometimes promoted as the first turbocharger used in a U.S. production car, and a display engine, now on exhibit in the Indy Hall of Fame museum has a plaque so indicating. However, turbochargers were offered on other cars (Corvair Monza and Olds Cutlass) as early as 1961. Pontiac's claim seems to be based on the fact that the 1980 Indy Pace Car was the first domestic car to have a turbo-charged engine as *standard* equipment.

Included with the turbo engine package was a

An off-center power blister on the hood was featured with a new turbocharged V-8. This 301-ci 210-horsepower engine was offered in the Formula and Trans Am and is shown here in a Formula with the extra-cost Formula Appearance package. With regular engines, the Formula used the 1979-style twin-scoop hood.

specific hood designed to accommodate the turbocharger. It had an off-center power bulge and carried "Turbo 4.9" lettering on the left-hand side. The new turbo-cast aluminum wheels could be ordered in silver or gold.

The turbo wheels were first available only for turbo-charged Formulas and Trans Ams and only when teamed with the WS6 performance package. However, this prevented California buyers from ordering them, as the engine wasn't available to them. But in November, the restrictions were modified to allow use of the wheels on California cars having the base 5.0-liter V-8 teamed with the WS6 option. With the exception of turbo wheel availability (in place of snowflake wheels at no extra cost), both the Special Performance package and the Formula Appearance package had the same ingredients as in 1979.

Priced at $7,179 on introduction date, the Trans Am was virtually an unchanged car except for new power teams and options. Available powerplants were the same as those offered for Formulas, but the 4.9-liter E/C was the standard federal engine. The 140 net horsepower version was a delete-option, giving the buyer a $180 credit. The same amount of credit was also extended to all California buyers who could have only the 5.0-liter V-8.

Priced noticeably higher, the black-and-gold Special Edition package returned with slightly gaudier striping and decals. Gold tape stripes highlighted the grille, airfoils, wheel lips, all windows, windshield, roof edges, air intakes, and the spoiler. The sail panel decals were larger and the front fender lettering was white. Also, the upper left-hand grille surround did not have lettering.

While the standard hood decal looked the same, a second turbo version was provided to go with the turbo hood bulge. The bird in the decal had narrower wings, a left-facing upturned head and larger flames emanating from its beak. At least five different types of engine call-out decals appeared on 1980 Trans Ams. They read "4.9," "4.9 T/A," "5.0 T/A," "Turbo 4.9" (early), and "Turbo T/A" (late).

The Indy Pace Car came only with the 4.9-liter turbo V-8, which excluded it from the California market. It was two-toned (white upper and gray lower) and had tri-color accent stripes. Oyster-colored vinyl bucket seats with oyster and black hobnail cloth inserts were standard. Birds were embroidered onto the door pads and center of the rear seat, and the exterior carried specific limited-edition decals including a super bird on the hood. There were white headlamp bezels, silver-tinted roof hatches, and red instrument panel lighting, too.

On its equipment list were the turbo hood, air conditioning, automatic transmission, power windows, 3.08:1 rear axle, custom seat belts, dual sport mirrors, leather Formula steering wheel, halogen headlights, pedal trim package, 64-amp

There were few changes in the 1980 Trans Am or the black Special Edition package shown here. The off-center hood blister indicates use of the 210-horsepower engine. Note the Turbo 4.9 lettering on the side of the power blister.

alternator, tinted glass, Special Performance package with 15x8-inch white turbo-cast wheels, 225/70R-15 white-letter tires, and front and rear power disc brakes.

For audio buffs, the Turbo Pace Car featured an electronically-tuned, digital-display AM/FM stereo radio with seek-and scan, cassette tape, and audio power booster. This car's price tag was $11,194.52. Amazingly, 14 extra-cost options were available for it.

Firebird VINs continued to follow basically the same system, with only letters used to designate engines. One change was the use of letters, instead of numbers, to indicate model year: an A in the sixth spot indicated 1980. The engine was coded in the fifth slot. The following chart lists engine codes for 1980 (note that there are no breakouts by transmission type because all 1980 Firebirds had automatics).

Code	Type	Displacement	Capacity	Source
A	V-6	231 ci	3.8 liter	Buick
W	V-8	301 ci	4.9 liter	Pontiac
Y	V-8	301 ci	4.9-liter E/C	Pontiac
T	V-8	301 ci	4.9-liter Turbo	Pontiac
G, H, S	V-8	305 ci	5.0 liter	Chevrolet

Early rust-outs, poor-quality paint, and loose bodies continued to trouble 1980 Firebird owners. There was also a recall to correct a steering problem: The new type of lower ball joints were not always sufficiently tightened at the connecting steering knuckle and were prone to separation.

Although magazine editors weren't road testing as many American cars as they had a few years earlier, they lined up to be behind the wheel of the new turbo-charged models. In October 1979, *Car and Driver* put a Turbo Trans Am through its paces and got a 0–60 time of 8.2 seconds and a 116-mile-per-hour top speed. The quarter-mile was covered in 16.7 seconds at 86 miles per hour.

A month later, *Motor Trend* got to wring out an Indy Pace Car and gave it good reviews on performance. However, when Pontiac expert Joe Oldham had a chance to test a new turbo, he wasn't quite as impressed.

"The harsh reality of the 1980 Turbo Trans Am was a mere shell of its former performance self," Oldham says in his book, *Supertuning Your Firebird Trans-Am.* "We had the distinct impression that we were driving a normally-aspirated 350 two-barrel powered Firebird of several years ago.

The Turbo Trans Am was selected for use as the official Pace Car for the 1980 Indy 500. A white Indy Pace Car replica was released to commemorate this honor. It had a multi-tone-gray, big bird hood decal and oyster and black Custom interior. A Turbo Trans Am was also selected as Daytona 500 Pace Car that season.

About 5,700 copies of the 1980 Trans Am Indy Pace Car were built. Collectors consider them very desirable today. Horizon Sportswear, Inc., of Madison Heights, Michigan, manufactured a special

Trans Am Pace Car jacket shown in this photo of a pace car parked by a historic Indianapolis race car. *Indianapolis Motor Speedway Corporation*

The regular Trans Am big bird hood decal, on the left, had the head facing to the right and a shorter wingspan. The decal on the right was designed for use on cars with the turbocharged engine and off-center hood blister. Part of the decal ran up and onto the bulged section of the hood.

A 1980 Trans Am turbo engine is on display at the Indianapolis Hall of Fame Museum, located in the speedway complex. A total of 23,422 Trans Ams and Formulas had this engine installed during 1980.

There is no special punch, no kick, no special feeling with the turbo. What you do get is small V-8 economy with medium V-8 performance."

In general, the 1980 Firebirds do not look like fast-appreciating cars at the current time. They are basically still used cars, with little special interest to collectors, unless an example with practically no miles on the odometer is found.

Some Pontiac enthusiasts do see a future for both turbo-powered Formulas and Trans Ams, and they say that the engine option probably warrants a one-star premium.

As for the Turbo Pace Car, it is a different story. This model combines a lot of things—limited-edition appeal, low production, pace car status, turbo engine, special appearance features, heavy options list, and luxury car characteristics—in one basically attractive package. So, despite the fact that its far from an all-out high-performance car, the 1980 Indy Pace Car looks like a highly collectible automobile that's in strong demand.

The following chart summarizes 1980 production totals:

Name	VIN	Engine	Code	Manual	Automatic	Total
Base Firebird	S87	—		—	—	29,811
Esprit	T87	—		—	—	17,277
Formula	V87	—		—	—	9,356
Trans Am Coupe	W87	4.9 liter	L37	—	14,866	14,866
		4.9-liter Turbo	LU8	—	5,753	5,753
		5.0 liter	LG4	—	1,635	1,635
Trans Am w/T-top	W87	4.9 liter	L37	—	8,692	8,692
		4.9-liter Turbo	LU8	—	7,176	7,176
		5.0 liter	LG4	—	896	896
Trans Am Special Edition	W87	4.9 liter	L37	—	72	72
Trans Am Special Edition w/T-top	W87	4.9 liter	L37	—	2,084	2,084
		4.9-liter Turbo	LU8	—	3,444	3,444
		5.0 liter	LG4	—	463	463
Trans Am Turbo Indy Pace Car*	X87	4.9 liter Turbo	LU8	—	5,700	5,700
Total						107,340

*Options available included UA1 heavy-duty battery, K30 cruise control, A90 remote-control deck lid release, C49 electric rear window defroster, K05 engine block heater, BS1 acoustical insulation package, TR9 lamp group, D34 visor vanity mirror, B84 white vinyl body side moldings, B93 door edge guard moldings, U75 power antenna, UA3 power door locks, N33 tilt steering wheel, and CD4 controlled-cycle windshield wipers.

1980 Firebird Options and Accessories

Sales Code	Description	Retail Price ($)
LD5	3.8-liter V-6 (S87, T87; n.a. w/VJ9)	n/c
L37*	4.9-liter V-8 S87, T87;	
	na w/VJ9; incl. chrome splitter	$ 180.00
	V87; n.a. w/VJ9	n/c
	W87; n.a. w/VJ9 less dual resonators	(180.00)
W72	4.9-liter E/C n.a. w/VJ9;	
	V87 incl. dual resonators	180.00
	na w/VJ9; W87	n/c
LU8	4.9-liter Turbo	
	(na w/VJ9; V87 incl. dual resonator)	530.00
LU8	4.9-liter Turbo (W87; n.a. w/VJ9)	350.00

Sales Code	Description	Retail Price ($)
LG4	5.0-liter V-8	
	Req. w/VJ9 and C60; S87, T87	195.00
	Req. w/VJ9 and C60; adds dual res.; V87	n/c
	Deleted std. chrome splitters; W87	(180.00)
MX1	Auto. trans. (S87, T87)	358.00
C60	Air conditioning	566.00
K81**	H-D 63-amp. alternator	
	w/o C49, C60	36.00
	w/C49, C60	n/c
W73	Yellow Bird	
	V87 only; w/velour hobnail trim	550.00

Sales Code	Description	Retail Price ($)
	V87 only; w/vinyl doeskin trim	505.00
W50	Formula Appearance pkg. (V87 only)	100.00
Y84	Special Edition	
	W87 only; w/o CC1	748.00
	W87 only; w/CC1	1,443.00
G80	Limited slip differential (all exc. W87)	68.00
UA1	HD-battery	23.00
AK1	Custom seat belts (exc. T87 w/o B18)	25.00
U35	Elec. quartz clock (w/o Rally gauges)	30.00
K30	Cruise control (req. w/MX1; w/V-6 req. w/JL2)	112.00
A90	Remote-control deck lid release	26.00
C49	Elec. rear window defrost.	107.00
VJ9	Calif. emissions	250.00
K05	Engine block heater	16.00
W63	Rally gauges (S87, T87, req. w/V-8 std. w/V87)	91.00
U17	Rally gauges	
	S87, T87; req. w/V-8	159.00
	V87; std. w/W87	68.00
A01	Tinted glass	68.00
CC1	Hatch roof; w/o Y84 (std. w/Y84)	695.00
TT5	Halogen headlights (hi-beam only)	27.00
D53	Hood decal w/o Y84 (std. w/Y84)	120.00
BS1	Added acoustical insulation (std. w/T87)	34.00
C95	Dome reading lamp	21.00
TR9	Lamp group	22.00
B37	Front & rear, color-keyed mats	27.00
D35	Sport OSRV mirrors (LH remote control; S87)	47.00
D34	Visor vanity mirror	7.00
B84	Color-keyed bodyside moldings	46.00
B93	Door edge guards	14.00
B83	Rocker panel molding (S87 only; std. w/T87; n.a. w/others)	20.00
B80	Scalp moldings (std. w/T87, W87 w/Y84)	26.00
B96	Wheel opening moldings (std. w/T87; n.a. w/others)	24.00
B85	Windowsill & rear hood moldings (std. w/T87, W87 w/Y84)	28.00
JL2	Power front disc/rear drum brakes (S87; T87; std. w/others)	81.00
J65	4-wheel power disc brakes	
	V87, W87 w/G80; w/o WS6	162.00
	V87, W87 w/G80; w/WS6	n/c
AU3	Power door locks	93.00
A31	Power windows	143.00
V02	Super cool radiator	
	w/C60; n.a. w/LU8	35.00
	w/o C60; n.a. w/LU8	64.00

Sales Code	Description	Retail Price ($)
U75	Power antenna	
	w/o opt. radios; n.a. w/UP6 radio	70.00
	w/opt. radio; n.a. w/UP6 radio	51.00
U83	Power AM/FM/CB tri-band antenna	
	w/o opt. radio	93.00
	w/opt. radio	74.00
U63	AM radio	97.00
UM1	AM radio w/integral 8-track stereo tape player	249.00
U69	AM/FM radio	153.00
UM7	AM/FM ETR radio w/seek-and-scan (incl. digital clock)	375.00
U58	AM/FM stereo radio	192.00
UN3	AM/FM stereo radio w/cassette stereo tape player	285.00
UM2	AM/FM stereo radio w/integral 8-track stereo tape player	272.00
UP6	AM/FM stereo w/40-channel CNB; incl. U83	525.00
UN9	Radio accommodation package	
	Std. w/opt. radio; w/o U75, U83	29.00
	Std. w/opt. radio; w/U75, U83	10.00
U80	Rear seat speaker (w/U63, U69 only)	20.00
UX6	Dual front speakers (w/U63, U69 only)	14.00
UP8	Dual front/rear speakers (w/U63, U69 only)	43.00
UQ1	Dual rear extended range speakers	
	w/U63, U69	68.00
	w/UM1/U58/UN3/UM2/UP6	25.00
	w/UM7 ETR radio	n/c

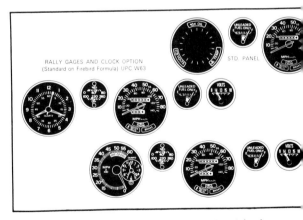

RALLY GAGES AND CLOCK OPTION
(Standard on Firebird Formula) UPC W63

STD. PANEL

The standard instrument-panel layout (top) had three slightly smaller round gauges. The Rally gauge cluster with clock (center) was standard on the Formula and optional on base Firebirds and Esprits. The Rally gauge cluster with tachometer and clock (bottom) was standard on Trans Ams, optional otherwise.

Sales Code	Description	Retail Price ($)
WS6	Special Appearance pkg.	
	V87 w/L37, LG4, LU8, W72	481.00
	W87; req. w/L37, LG4, LU8, W72; w/o Y84	481.00
	W87; req. w/L37, LG4, LU8, W72; w/Y84	281.00
D80	Rear deck spoiler	
	S87, T87	62.00
	V87 w/o W50	62.00
	V87 w/W50, W87	n/c

Sales Code	Description	Retail Price ($)
NK3	Formula steering wheel	
	S87	74.00
	T87 w/o W73	52.00
	T87 w/W73, V87, W87	n/c
N30	Luxury cushion steering wheel (S87)	22.00
N33	Tilt steering wheel	81.00
D98	Vinyl tape stripes (S87 and T87 w/o W50 or W73 only)	58.00

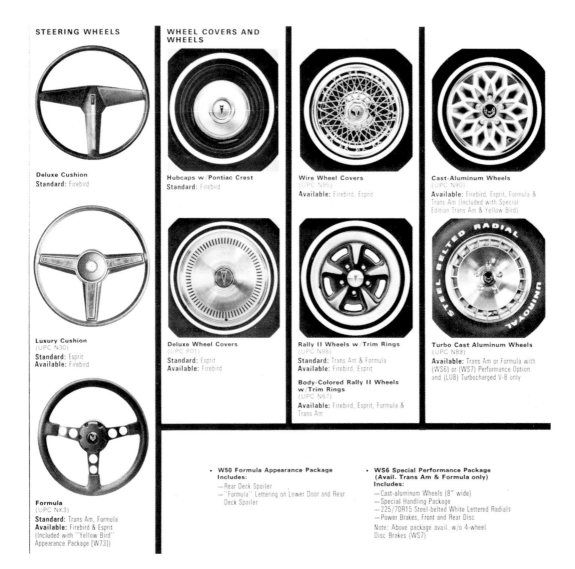

STEERING WHEELS

Deluxe Cushion
Standard: Firebird

Luxury Cushion
(UPC N30)
Standard: Esprit
Available: Firebird

Formula
(UPC NK3)
Standard: Trans Am, Formula
Available: Firebird & Esprit
(Included with "Yellow Bird" Appearance Package [W73])

WHEEL COVERS AND WHEELS

Hubcaps w/Pontiac Crest
Standard: Firebird

Deluxe Wheel Covers
(UPC P01)
Standard: Esprit
Available: Firebird

Wire Wheel Covers
(UPC N95)
Available: Firebird, Esprit

Rally II Wheels w/Trim Rings
(UPC N98)
Standard: Trans Am & Formula
Available: Firebird, Esprit

Body-Colored Rally II Wheels w/Trim Rings
(UPC N67)
Available: Firebird, Esprit, Formula & Trans Am

Cast-Aluminum Wheels
(UPC N90)
Available: Firebird, Esprit, Formula & Trans Am (Included with Special Edition Trans Am & Yellow Bird)

Turbo Cast Aluminum Wheels
(UPC N89)
Available: Trans Am or Formula with (WS6) or (WS7) Performance Option and (LU8) Turbocharged V-8 only

• **W50 Formula Appearance Package**
Includes:
— Rear Deck Spoiler
— "Formula" Lettering on Lower Door and Rear Deck Spoiler

• **WS6 Special Performance Package**
(Avail. Trans Am & Formula only)
Includes:
— Cast-aluminum Wheels (8" wide)
— Special Handling Package
— 225/70R15 Steel-belted White Lettered Radials
— Power Brakes, Front and Rear Disc
Note: Above package avail. w/o 4-wheel Disc Brakes (WS7)

Three steering wheels (left) and six styles of hubcaps, wheel covers, and wheel rims were available in 1980. The Alliance turbo cast-aluminum wheels were used as standard equipment on Indy pace cars.

Sales Code	Description	Retail Price ($)
B18	Custom trim group	
	V87, W87 w/vinyl doeskin trim	142.00
	V87, W87 w/hobnail cloth trim	187.00
	T87 w/hobnail cloth trim w/o W73	45.00
P01	Deluxe wheel covers (S87)	45.00
N95	Wire wheel covers	
	S87	171.00
	T87	126.00
N18	Wheel cover locking package (W/N95 only)	35.00
N90	Cast aluminum wheels (4)	
	S87	336.00
	T87 w/o W73	291.00
N98	Rally II wheels & 4 trim rings	
	S87 w/N65	136.00
	S87 w/o N65	158.00
	T87 w/N65	91.00
	T87 w/o N65	113.00
	V87, W87	n/c
N67	Body-color Rally II wheels & 4 trim rings	
	S87 w/N65	136.00
	S87 w/o N65	158.00
	T87 w/N65	91.00
	T87 w/o N65	113.00
	V87, W87	n/c
—	Turbo cast wheels (req. w/LU8 turbo V-8)	_

Sales Code	Description	Retail Price ($)
UR4	Turbo boost gauge (req. w/LU8 turbo V-8)	40.00
CD4	Controlled-cycle windshield wipers	41.00
N65	Stow away spare	n/c
QJU	205/75R15 black steel-belted radial tires (S87, T87)	n/c
QJW	205/75R15 whitewall steel-belted radials	
	S87, T87 w/o N65	62.00
	S87, T87 w/N65	50.00
QMC	205/75R15 white-letter steel belted radials	
	S87, T87 w/o N65	80.00
	S87, T87 w/N65	64.00
QGR	205/75R15 white-letter steel-belted radials	
	w/RTS (V87, W87 w/o WS6)	68.00

*L37 engine for (S87) Firebird and (T87) Esprit increased to $225 around April 1980.

**70-amp. alternator was released in April at following prices: w/o C49 and w/o C60, $51.00; w/C49 or w/C60, 15.00; w/both C49 and C60, no charge. The 70-amp alternator was not available w/LD5 or LG4 engines.

_No price was listed for the turbo cast aluminum wheels. They could be installed in place of Rally II wheels when teamed w/specific other options.

Chapter 15

1981

★	Firebird V-6, V-8
★	Esprit V-6, V-8
★	Formula
★★★	Turbo Formula
★★	Trans Am
★★	Trans Am Special Edition
★★★	Trans Am Special Edition coupe w/5.0 Liter
★★★	Turbo Trans Am
★★★★	Turbo Trans Am Special Edition
★★★★★	NASCAR (Daytona) Turbo Pace Car

Externally, there were only small changes in Firebirds for model year 1981. A white bird emblem was placed on the fuel filter door between the taillights. New in front was a black-finished grille with argent silver accents. Overall length was listed as 198.1 inches (1.3 inches longer than the 1980 specification). There were new exterior colors, with choices for Trans Ams expanded to 13 hues.

Inside all models, a floor console up front was again standard equipment. A few common things like a cigarette lighter and front ashtray were specifically mentioned as standard equipment to make the features list look longer and more impressive. A silver, doeskin vinyl interior selection was added and oyster-colored vinyl was dropped. A new optional combination featuring Pimlico cloth bucket seats with Durand cloth bolsters came in a new beige combination.

Engineering refinements that were standard on all models included low-drag front disc brakes, a quick-take-up brake master cylinder, an early fuel evaporation system, a new lightweight Delco Freedom battery with side terminals, and GM's Computer Command Control system.

Computer Command Control was a high-technology system that utilized an on board solid-state microprocessor to create an engine that "thought." The result was a line of cars offering the highest fuel economy and lowest emissions levels in GM's history.

Because of the system, Pontiac was able to shuffle its Firebird power team line-up to introduce new engines and bring back some features that customers had been asking for. The 4.3-liter V-8, which had been put on hold in 1980, was initially made standard equipment for 1981 Formulas. An added option for Trans Ams and Formulas was a Chevrolet-built 5.0-liter V-8 teamed with four-speed manual transmission. An automatic transmission was standard in Esprits and a three-speed manual gearbox was returned for use in base Firebirds, although it could only be had in virtually optionless cars.

Modified 1981 options included a cruise control system with resume-speed function, an improved ETR sound system with digital read-out display, a revised 4.9-liter E/C engine, a redesigned Formula Appearance package, an improved lamp group, a different cassette stereo tape player, an updated WS6 Special Performance package, and two-tone hood decals (in five color combinations) for Trans Ams. The turbo engine package, offered in Formulas and Trans Ams, incorporated a new turbo-boost light display in the special, off-center hood power bulge.

A movie titled Smokey and the Bandit, starring actor Burt Reynolds, had given national exposure to a Turbo Trans Am, and Pontiac capitalized on its popularity. Reynolds, in appropriate "Bandit" attire, appeared on the centerfold of the 1981 Pontiac sales catalog.

In the summer, a sequel called Smokey and the Bandit II inspired a New Jersey company to market an interesting Trans Am Bandit model featuring such additions as Recaro seats, a

Blaupunkt stereo, Goodyear Eagle GT tires, an Escort radar detector, a Doug Nash five-speed transmission, and a 462-ci, 380 net horsepower engine. Priced at $30,000, the Bandit was to be released in a limited run of 200 copies. There were also a number of aftermarket Firebird convertible conversions offered by various specialty companies that year, but neither the Bandit nor these ragtops were factory-built cars and, thus, they are beyond the general scope of this book.

The factory did release another midyear pace car package, however. The effort was tied into the Daytona 500 stock car race run in February, but it is generally referred to as the NASCAR Turbo Pace Car. This car looked similar to the previous season's Indy 500 package and was built in a limited run of 2,000 copies.

Perhaps the most significant thing about the 1981 factory models, from the collector's standpoint, is the fact that they represented the last of the second-generation Firebirds. A brand new F-car was in the works for 1982—and it was needed. After 12 years in the marketplace, the Firebird had turned into a big, old-fashioned gas hog. When the last Pontiac-built V-8 left the engine assembly plant in March 1981, the writing was on the wall. A production drop to just under 71,000 total units was registered for the year, as if to toll the final death knell.

Pontiac was suddenly stressing V-6 economy in the base Firebird, which retailed for a very uneconomical $7,154.59. It had the new exterior

features and standard equipment outlined above, but looked much the same as the 1980 edition.

The base engine nationwide was the 3.8-liter V-6 with five fewer horsepower. The three-speed manual transmission was available only with this engine and floor-mounted shift controls, and came with numerous restrictions. You could not get it on cars with California emissions, air conditioning, added acoustical insulation, front and rear floor mats (B37), rocker panel moldings, rear deck spoiler, Rally II wheels, or any radio. It was also not available in Esprits teamed with the W63 Rally gauge cluster. In fact, a check of Ward's Automotive Yearbook 1982 shows that 93.8 percent of all 1981 Firebirds had automatic transmissions and 5.8 percent had a four-speed manual. This means that 0.004 percent of all 1981 Firebirds (or 283 cars) left the factory with the three-speed gearbox.

Base Firebirds were also available with the V-6 and automatic transmission, plus two V-8 options. The first was the 4.3-liter engine with two-barrel carburetion, 8.3:1 compression and 120 net horsepower at 4,000 rpm. This option required the automatic transmission and air conditioning. Also available was the 4.9-liter E/C engine, which also required an automatic transmission and included chrome splitter tail pipe extensions. Sadly, it had five fewer horsepower than in 1980.

Carrying a base price of $8,020.59, the 1981 Esprit had the same basic package of luxury

Primary changes for the 1981 base Firebird coupe were invisible to the naked eye. It remained the only Firebird to come standard with a single, outside rearview mirror. This car also has the standard vinyl interior.

The 1981 Esprit was the last Firebird (through 1986 at least) to feature this ten-year-old model name. The Esprit featured the Custom interior option with pleating on seats running widthwise instead of lengthwise. The rear spoiler and cast-aluminum wheels shown here were extra-cost options.

extras. A chrome script reading Esprit was below the Firebird lettering on the front fenders, and color-keyed door handle inserts remained the model's trademark. The Esprit came with the same power team available for base Firebirds at the same prices.

When introduced in the fall of 1980, the Formula had a base price of $8,143.23 and the 4.3-liter V-8 as standard equipment. In 1981, a rear deck lid spoiler was included as regular equipment. The new two-tone Formula Appearance package was something else. It featured a lower-body perimeter finish in a contrasting hue with Formula graphics and a tape stripe completely around the perimeter. The next section of the body sides, up to the lower beltline, was in another color. The upper body (including roof) was in a third color. When the turbo package was added, Turbo Formula lettering appeared above the taillights, and the turbo blister hood was used. Such a car, in dark maroon, charcoal, and silver finish was depicted in the year's sales catalog.

A major midyear change for Formulas was a revision in the standard powerplant. As listed in a 1981 Pontiac Selling Facts booklet updated in April 1981, the 3.8-liter V-6 became the standard engine for this model. This change, which did not include any price revisions, made the late-season Formula the only Firebird to offer all five powerplants available for the season.

Most likely, one reason for the change was that Pontiac had stopped building V-8s in March. Engines were still available from inventory, but it didn't make sense to promote their use as standard equipment. When the V-6 was made the base engine, the 4.3-liter V-8 became a $50 option. When installed in Formulas, both before and after the change, the 4.3 liter had to be teamed with W63 Rally gauge cluster option, automatic transmission, and air conditioning.

The 4.9-liter E/C engine was $215 extra in Formulas throughout the year. Ordering this option required the use of an automatic transmission and W63 Rally gauges, and included chrome splitters. Other specifications were the same as for base Firebirds.

The Formulas also shared two other options with the Trans Am. The first was the Chevy-built 5.0-liter V-8 with four-barrel carburetion. This

A new, Pontiac-built 265-ci two-barrel V-8 was standard in the 1981 Formula Firebird coupe.

Shown here is a car with the W50 Formula Appearance package. *Pontiac Motor Division*

was about the snappiest non-turbo option available and had 8.6:1 compression and 145 net horsepower at 4,000 rpm. It required the four-speed manual transmission and (on Formulas) the W63 Rally gauges, and it included dual resonators and tail pipes. The engine was $75 extra in Formulas, but a delete-option in Trans Ams.

For $652, Formula buyers could also order the 4.9-liter turbo, which had slightly lower compression (7.5:1) and horsepower (200 net horsepower at 4,000 rpm) than in the previous season. The turbo required air conditioning, an automatic transmission, front and rear power disc brakes, the limited-slip axle, and W63 Rally gauges, but included chrome exhaust splitters, turbo-boost gauge, and the turbo blister hood.

Listing for $8,670.23, the 1981 Trans Am had nothing that was earth-shakingly new. Its base engine was the 4.9-liter E/C V-8 (also called the "Trans Am 4.9 liter"). The 5.0-liter V-8 was available as a delete-option (giving a $140 credit), and it replaced chrome splitters with dual resonators and pipes. At $437 extra, the turbo 4.9-liter engine was the top option for the base Trans Am. It added the special turbo blister hood and turbo-boost gauges.

Merchandised as a separate model with a $12,623.23 base price, the Turbo Trans Am Special Edition included all standard Trans Am features plus the 4.9-liter turbo package, power windows, Custom trim, hatch roof, hood decal, ETR signal-seeking, AM/FM stereo radio with cassette, and the WS6 Special Performance package.

There was still a non-turbo Special Edition package available without ($779) or with ($1,516) the hatch roof panel treatment. Most cars getting this option had the T-tops added; in fact, the solid-top versions were rare—only 121 were built in total, and only 15 of them came with the 5.0-liter V-8. (Because of this low production total, I have given a higher star rating to this model.) The ingredients of the regular Special Edition package were the same as for 1980.

The NASCAR Turbo Pace Car, released around February, was built in a limited run of 2,000 copies with a $12,244 base price. It looked similar to the previous season's Indy Pace Car with black (instead of charcoal) accents. The 1981-style hood decal was featured along with Recaro bucket seats (black with red inserts and carpeting), turbo lights, hatch roof, and white turbo cast aluminum wheels. Pace Car decals were shipped inside the cars, for dealer installation only, if the buyer preferred the graphics.

Pontiac's VIN system was changed in 1981. Numbers were still found on the top left-hand

This 1981 Trans Am has the off-center hood bulge indicating use of the 4.9-liter (301-ci) turbocharged V-8. *Pontiac Motor Division*

Actor Burt Reynolds drove a black-and-gold Turbo Trans Am to Hollywood stardom in the film *Smokey & the Bandit*. The car, a T-top coupe with many custom touches, was called the Bandit. *Pontiac Motor Division*

surface of the instrument panel and visible through the windshield. There were 17 characters. The first three identified the manufacturer, make, and type of vehicle. The fourth letter identified the restraint system. The fifth, sixth, and seventh characters identified the series and body style. The eighth identified the engine. The ninth character was a check symbol. The tenth was a letter designating model year. The eleventh character told which factory the car was made in. The last six digits were the sequential production number.

A typical Firebird number looked like this: ( )G2( )( )( )AB( )100001 and up. In this example, the G2 designates Pontiac Firebird, the A is the check symbol, and the B indicates 1981. The empty space in front of the check symbol would be stamped with one of the engine codes.

Code	Type	Displacement	Capacity	Source
A	V-6	231 ci	3.8 liter	Buick
S	V-8	265 ci	4.3 liter	Pontiac
W		301 ci	4.9 liter E/C	Pontiac
T			4.9 Turbo	Pontiac
G, H, S		305 ci	5.0 liter	Chevrolet or GM of Canada

There were many improvements in 1981 Firebirds that helped to enhance reliability and owner satisfaction levels. Cushion mounting for the subframe helped eliminate rattles. Pontiac claimed that the cushion body mounting system also effectively helped to isolate the passenger area from road noise and vibration. Extensive anticorrosion measures were being used, including the attachment of inner fender panels, front and rear, to reduce rust-out in these areas. Improved double-panel construction was featured in the doors, hood, and deck lid, and the hood and deck lid were counterbalanced.

Such measures did not totally cure past problems, however. When Motor Trend tested one of the Bandit Trans Ams, in its June 1981 issue, the magazine criticized its space utilization and noted, "a chorus of creaks and groans (that) says all that must be said about chassis rigidity." In short, the 1981 bodies were better, but far from perfect.

Under the hood were more improvements, thanks to Computer Command Control, which resulted in cleaner-running, more fuel efficient engines. With a six, the Firebird was now rated at 19 miles per gallon by the EPA. The 4.3-liter V-8 had an 18 miles per gallon rating and the 4.9-liter E/C engine was good for 15 miles per gallon. There were no EPA estimates for the 4.9-liter turbo nor the 5.0-liter V-8, although the latter powerplant was rated for 16 miles per gallon in the similar Camaro.

Two different magazines tested the Turbo Trans Am for performance. Road Test recorded an 8.7 second 0–60 time and a quarter-mile run of 16 seconds at 85.9 miles per hour. Road & Track clocked 9.8 seconds for 0–60, but didn't specify quarter-mile performance. Road Test noted overheating problems after several runs down the quarter-mile.

As far as collectibility goes, the 1981 Firebirds offer several desirable option packages and three option and engine combinations that are particularly hard to find. The limited edition NASCAR Turbo Pace Car also had definite collector appeal and is one of the less frequently seen specialty models. In addition, with overall low production and "last big Firebird" status, 1981 models should receive growing interest.

Total assemblies of 1981 Firebirds were recorded as 70,899 units. Available breakouts by option packages and powerplants looked like this:

Name	VIN	Engine	Manual	Automatic	Total
Base Firebird	S87	—	—	—	20,541
Esprit	T87	—	—	—	10,938
Formula	V87	—	—	—	5,927
Trans Am Coupe	W87	4.9-liter E/C	—	5,087	5,087
		4.9-liter Turbo	—	3,851	3,851
		5.0 liter	2,866	—	2,866
Trans Am w/T-top	W87	4.9-liter E/C	—	4,589	4,589
		4.9-liter Turbo	—	6,612	6,612
		5.0 liter	3,225	—	3,225
Special Edition Coupe	X87	4.9-liter E/C	—	41	41
		4.9-liter Turbo	—	65	65
		5.0 liter	15	—	15
Special Edition w/T-top	X87	4.9-liter E/C	—	1,160	1,160
		4.9-liter Turbo	—	3,050	3,050
		5.0 liter	932	—	932
NASCAR Turbo Indy Pace Car	X87	4.9-liter Turbo	—	2,000	2,000
Total					70,899

1981 Firebird Options and Accessories

Sales Code	Description	Retail Price ($)
LD5	3.8-liter V-6 (S87, T87, V87 after midyear)	n/c
LS5	4.3-liter V-8 2V (S87, T87; std. in V87 until midyear)	$ 50.00
L37	4.9-liter E/C V-8 4V (S87, T87, V87; std. w/W87; req. w/MX1, C60 & includes chrome splitters; req. w/W63 on V87)	215.00
LU8	4.9-liter turbo V-8 4V (incl. turbo-boost gauge & on V87, chrome splitters; also req. w/MX1, C60, J65, G80; also incl. dual resonator/tail pipes; req. w/W63 on V87)	
	in V87	652.00
	in W87	437.00
	in X87	n/c
LG4	5.0-liter V-8 4V (req. w/MM4; incl. dual resonator/tail pipes; also req. w/W63 in V87)	
	in V87	75.00
	in W87	(140.00)
MM3	3-speed man. trans. (incl. floor shift; avail only w/LD5; not avail. w/VJ9, C60, BS1, B37, B83, D80, N98 or any radios; not avail. in T87; avail. in V87 w/W63 after midyear)	n/c
MM4	4-speed man. trans. (incl. floor shift; req. w/LG4; V87, W87)	n/c
MX1	Auto trans. (req. w/LS5, L37, LU8 V-8)	
	in S87	370.00
	in T87, V87, W87, X87	n/c
C60	Custom air condition (req. w/X87)	600.00
K73	H-D 70-amp. alternator	51.00

Sales Code	Description	Retail Price ($)
W50	Formula Appearance pkg. (V87 only)	212.00
Y84	Special Edition Appearance pkg.	
	W87; X87; w/o CC1	779.00
	W87; X87; w/CC1	1,516.00
G80	Limited-slip axle	
	S87, T87	71.13
	V87, W87 w/o WS6	71.00
	V87; W87 w/WS6; w/X87 also	n/c
UA1	H-D battery	22.00
AK1	Custom seat belts (all exc. T87, X87 w/o B18; incl w/B18)	26.00
U35	Electric quartz clock (na w/UM6, UM7; std. w/W63, U17)	30.00
K35	Cruise control (req. w/MX1)	145.00
A90	Electric rear window defroster	115.00
VJ9	California emissions	46.00
K05	Engine block heater	17.00
W63	Rally gauges, clock & trip odometer (na w/UM7; req. w/V87 w/V-8 engines; std. in W87, X87)	95.00
U17	Rally gauges, clock, trip odometer and tach. (std. w/W87, X87)	166.00
A01	Soft-Ray glass, all windows	82.00
CC1	Hatch roof (all exc. X87 w/o Y84; std. w/Y84)	737.00
TT5	Halogen headlights	29.00
D53	Hood decal (W87 w/o Y84; std. w/Y84, X87)	125.00
BS1	Added acoustical insulation (na w/MM3, X87; std. w/T87)	36.00
C95	Dome reading lamp	21.00

Sales Code	Description	Retail Price ($)
TR9	Lamp group (incl. tone generator; glovebox; Instrument panel courtesy & luggage)	35.00
B37	Front and rear floor mats (na w/LU8 in W87; Req. w/MX1 w/LD5)	25.00
D35	OSRV sport mirror (S87; std. w/T87, V87, W87, X87)	51.00
D34	Visor vanity mirror (RH)	7.00
B84	Body side moldings (vinyl; color-keyed)	44.00
B93	Door edge guard moldings	14.00
B83	Rocker panel moldings (S87, T87 only; n.a. w/MM3; n/c T87)	23.00
B80	Roof drip moldings (na X87; std. w/T87, Y84)	27.00
B96	Wheel opening moldings (S87, T87 only; std. w/T87)	26.00
B85	Window sill & rear hood edge moldings (std. w/T87, Y84; n.a. w/X87)	30.00
J65	Front and rear power disc brakes (avail. on W87, V87 w/G80 only; n/c w/WS6, X87)	167.00
A31	Power windows (all exc. X87; std. w/X87)	152.00
AU3	Power door locks	99.00
V02	Super cool radiator w/o C60; incl. w/LU8; n.a. w/LG4	67.00
	w/C60; incl. w/LU8; n.a. w/LG4	37.00
U75	Power antenna (na w/UP6; $20 less w/opt. radio)	70.00
U83	Power antenna (AM/FM/CB tri-band; incl. w/UP6;$20 less w/opt. radio)	92.00
U63	AM radio	90.00
UM1	AM radio w/integral 8-track stereo tape player	231.00
U69	AM/FM radio	142.00
UM7	AM/FM ETR stereo radio w/seek-and-scan, incl. digital	402.00
U58	AM/FM stereo radio	178.00
UM6	AM/FM ETR stereo radio w/tape, clock, digital (std. w/X87)	555.00
UN3	AM/FM ETR stereo radio w/tape, incl. UQ1 speakers	289.00
UM2	AM/FM stereo radio w/integral 8-track stereo player	252.00
UP6	AM/FM/CB stereo radio (incl. U83; $65 credit in X87)	490.00
UN9	Radio accommodation package	29.00
U80	Dual rear speakers (only w/U63 or U69 monaural radios)	20.00
UX6	Dual front speakers (only w/U63 or U69 monaural radios)	15.00
UP8	Dual front and rear speakers (only w/U63, U69, std. w/UM1, UM2)	42.00
UQ1	Dual rear extended-range speakers w/U63, U69; incl. front duals	67.00

Sales Code	Description	Retail Price ($)
	w/UM1, U58, UM2, UP6; std. w/UM6, UM7, UN3	25.00
UQ3	Audio power booster ($84 w/UN3, w/UM1, U58, UM2, UP6; included w/ UQ1)	109.00
A51	Bucket seats (S87, V87, W87 w/Sparta cloth trim)	28.00
WS6	Special Performance pkg. V87, W87 w/o Y84	580.00
	W87 w/Y84	372.00
D80	Rear deck lid spoiler (S87, T87; incl. w/V87, W87, X87)	64.00
NK3	Formula steering wheel S87	78.00
	T87; included w/V87, W87, X87	55.00
N30	Luxury cushion steering wheel (included w/T87; S87)	23.00
N33	Tilt steering wheel	88.00
D98	Vinyl tape stripes (na W87, X87 or V87 w/W50)	61.00
B18	Custom trim option T87; velour Pimlico cloth	48.00
	V878; W87; doeskin vinyl trim	148.00
	V87, W87; Pimlico cloth trim	196.00
P01	Deluxe wheel covers (S87; std. w/T87)	49.00
N95	Wire wheel covers S87	179.00
	T87	130.00
N18	Wire wheel cover locks	36.00
N90	Turbo-cast aluminum wheels na w/WS6, X87	351.00
	T87; n.a. w/WS6, X87	302.00
	V87; W87 w/o Y84 std. w/Y84	208.00
N98	Silver Rally II wheels and 4 trim rings (req. w/MX1 w/LD5) w/S87	143.00
	w/T87	94.00
	w/V87 and W87	n/c
N96	Body-color Rally II wheels & 4 trim rings (req. w/MX1 w/LD5) S87	143.00
	T87	94.00
	V87, W87	n/c
CD4	Controlled-cycle windshield wipers	44.00
QJW	205/75R15 whitewall steel-belted radial tires (S87, T87)	56.00
QMC	205/75R15 white-letter steel-belted radial tires (S87, T87)	74.00
QQR	225/70R15 white-letter steel-belted radial tires (V87, W87 w/o WS6)	78.00

In this publicity photo are examples of the first- and last-year models to have Pontiac V-8s: a 1967 Firebird 400 convertible on the bottom of the transporter and a 1981 Trans Am with four-barrel V-8 (note shaker-type hood scoop) on top of the transporter. Pontiac Motor Division

Chapter 16

1982, 1983, 1984, 1985, 1986

1982	
★	First-level Firebird Coupe
★★	Firebird S/E Coupe
★★★	Trans Am Coupe

1983	
★	First-level Firebird Coupe
★★	Firebird S/E Coupe
★★	Trans Am Coupe
★★★★	Daytona 500 Silver Anniversary T-top
★★★	Special Edition Recaro Trans Am Coupe
★★★	Trans Am Coupe (with HO V-8)

1984	
★	First-level Firebird Coupe
★★	Firebird S/E Coupe
★★	Trans Am Coupe
★★★	Special Edition Recaro Trans Am Coupe
★★★	Trans Am Coupe (with HO V-8)
★★★★	Aero Trans Am Coupe (with HO V-8)
★★★★★	15th Anniversary Trans Am T-top

1985	
★	First-level Coupe
★★	Firebird S/E Coupe
★★★	Trans Am Coupe
★★★★	Trans Am Coupe (with HO V-8)
★★★★	Trans Am Coupe (with TPI V-8)

1986	
★★	First-level Coupe
★★★	Firebird S/E Coupe
★★★	Trans Am Coupe
★★★★	Trans Am Coupe (with TPI V-8)

With sleeker looks, less bulkiness, space-age engineering, and a generally more exciting overall image, the early third-generation Firebirds came in three distinct models, each having a specific identity. The basic Firebird—usually described as the first-level model—was officially considered a compact car. The S/E was the next step up in features and price, and it represented a luxury touring model. Positioned just slightly ($34) above the S/E was the Trans Am, which stressed a performance image instead of the S/E's luxury look.

This line-up of models and the general character of the three cars did not change much between 1982 and 1986, making it sensible to treat the cars as one group. There have, however, been some interesting refinements from year to year.

Are these Firebirds collector cars? At this point, the answer is no. They aren't old enough—and haven't been researched enough—to be categorized as rare or special-interest models. Surely, collector interest in at least some of them will evolve as the years go by.

This buyer's guide surveys the late-model Firebirds in terms of basic product offerings, strong points, weak points, and applicable used-car values. Along the way, it attempts to touch on some points that may predict their appreciation potential and possible future appeal as collector cars.

Since all of the 1982–86 Firebirds are somewhat similar in areas like passenger comfort, convenience, driver features, workmanship, and durability, I will provide a general rundown of strong points and weak points in each category at the end of this chapter. However, there were noticeable changes in performance and handling from year to year, which are covered separately.

1982

When the all-new Firebirds arrived in January 1982, they were conventional front-engine, rear-drive vehicles with unit-body construction. The front suspension was a modified version of the MacPherson strut system. A torque-arm rear suspension was featured to positively locate the rear axle assembly.

Downsizing became part of the Firebird story. The new models were poised on a 101-inch wheelbase and measured 189.8 inches end to end. All three stood only 49.8 inches tall and had an overall width of 72 inches. Front and rear tread measured 60.7 and 61.6 inches, respectively. Weight in the base model was reduced by over one-quarter ton to 2,858 pounds.

Pontiac described the wedge-shaped Firebird body as having "a sabre-like nose and rakish tail." Soft S curves were seen over the fenders and a subtle "bone line" ran through the middle. The front tapered to an ultralow nose with split grilles housed inside air slots. Parking lights peered out of two slots above the grilles and below the electrically operated, hidden quartz-halogen headlights. A 62-degree windshield angle further enhanced the aerodynamics.

Body style availability was limited to a two-door hatchback coupe with a large, contoured, frameless, all-glass hatch. Usable cargo area, with rear seatback folded, was a huge 30.9 cubic feet. In wind tunnel tests, the three models produced

These Rally V full wheel covers, bodyside moldings, and white-letter tires were extra-cost options. Milwaukee Zone Pontiac dealers brought about 100 of the 1982 Firebirds to this promotional drive-away held in conjunction with a Trans Am Territory at the Road America race course in Elkhart Lake, Wisconsin.

drag coefficient numbers of .333 (first-level), .335 (S/E), and .323 (Trans Am).

The first-level (base) Firebird, selling for $7,155 in standard form, was clean and simple in overall appearance. Then came the S/E ($8,021), which was done in a subtle tone-on-tone color concept to give it a sophisticated look. It carried S/E badges on its sail panels and shared full-width, black-lens taillights with the Trans Am. Priced at $8,143, the Trans Am was at the top of the line. It featured a black-accented high-tech "structural" look and a functional rear decklid spoiler. A bird emblem appeared on the sail panels.

Standard equipment for the first-level included a 2.5-liter four-cylinder engine with electronic fuel injection (EFI), a four-speed manual transmission, power steering, power brakes, a center console, full carpeting, reclining front bucket seats, a fold-down rear seatback, dual outside mirrors, hidden headlights, and a Formula steering wheel.

To this, the S/E substituted a 2.8-liter V-6, black-finish exterior accents, color-keyed bodyside moldings, a lockable fuel filler door, sport

Basically intended to replace the Esprit in concept, the all-new Firebird S/E included matte-black body highlights among its standard equipment.

A Viscount cloth Custom trim interior was standard in the Firebird S/E and optional in all other models. Reclining bucket seats and a new type of Formula steering wheel were standard in all 1982 Firebird car lines. Doeskin vinyl or Parella cloth Luxury interiors were a no-cost addition for the S/E.

Leather seats with vinyl bolsters were optionally available for all of the new 1982 Firebirds. The price for this particular version of the B20 Luxury trim interior group was $844 for first-level and Trans Am models, and $545 for the Firebird S/E.

mirrors (with remote control for the left-hand unit), a rear window wiper/washer, body-color turbo-cast aluminum wheels, extra acoustical insulation, an electric hatch release, full instrumentation, Viscount cloth upholstery, special suspension, and a luxury interior with door map pockets.

To set it apart from the pack, the Trans Am added the following to the first-level equipment list: a 5.0-liter V-8, black-finished exterior accents, front fender air extractors, wheel opening flares, sport mirrors (left-hand remote control), a rear spoiler, turbo-cast aluminum wheels, full instrumentation, and special suspension. A hot option, available for the Trans Am only, was a 5.0-liter, dual throttle-body-injected V-8 with fresh-air induction.

Coding used to identify these cars was a bit simpler than in the past. The first-level model was designated S87, the S/E model was designated X87, and the Trans Am was designed W87. These codes appeared in the same 17-symbol VIN sequence used the previous year. The tenth symbol was changed to a C for 1982 (third letter in the alphabet for third year of the new, Government-mandated VIN system), and the engine codes for Firebirds were as follows:

Engine	Code
2.5-liter four	V, 1, 9, 5, F, R, 2
2.8-liter V-6	7, X
5.0-liter V-8	7

Public acceptance of the new models was very positive. In addition, magazine testers gave the cars basically high ratings. The weak points were economy and noise level, which were generally classified as fair. There was a marked improvement in workmanship, and performance characteristics were very good to excellent.

Model year production climbed to 116,364 total units in the abbreviated eight-month selling season. V-8 engines were used in 64,116 of the 1982 Firebirds, as the bulk of sales went to the sporty Trans Am. A total of 34,444 had V-6 engines, and only 17,804 units utilized the base four-cylinder power-plant. A total of 95,418 new Firebirds featured an automatic transmission; 20,946 had four-speed gearboxes. Air conditioning was a popular accessory, having been installed in 102,051 of these cars.

Road & Track (September 1982) tested a new Trans Am with the 5.0-liter throttle-body-injected V-8 and concluded that it was "larger

The redesigned 1982 full-length front console was standard in all models and integral with the dash. Hood-scoop opening visible through windshield reveals this to be a Trans Am. The Viscount seats shown were $28 extra, but the Rally gauge cluster—including tach, but not clock—was standard.

Trans Am featured full-width taillights hidden behind a black, opaque, ribbed lens. Lockable gas-filler door was optional on first-level and Trans Am, but standard with S/E. Endura rear bumper was mainly body colored, with contrasting finish on lower portion. Notice how new spoiler curved down at either end. A rear spoiler was standard with S/Es, optional on all other models.

and heavier than we'd like for these early Eighties [but] nonetheless a dramatic improvement on [its] predecessors." The magazine obtained a 0–60-mile-per-hour time of 9.2 seconds. It took 17 seconds to cover the quarter-mile from a standing stop, with a terminal speed of 80.5 miles per hour.

As far as handling, *Road & Track* compared the Trans Am to the similar, all-new Camaro and found the Pontiac setup had a slightly softer and more European style than the Camaro, which had a street-machine style. It was noted, however, that three of four editors preferred the Trans Am to its Chevrolet counterpart and that both cars exhibited excellent, though different, handling and braking characteristics.

1982 Firebird Options and Accessories

UPC Code	Description	Retail Price ($)
LQ9	2.5-liter 151-ci EFI L-4	
	Firebird	n/c
	Firebird S/E (na w/WS6) (credit)	(125.00)
LC1	2.8-liter 173-ci V-6 2-bbl	
	Firebird	125.00
	Firebird S/E	n/c
LG4	5.0-liter 305-ci V-8 4-bbl	
	Firebird (na w/MM4)	295.00
	Firebird S/E	
	w/MX1; w/o WS6	170.00
	w/MM4 or WS6 (incl. dual exhaust)	195.00
	Trans Am (incl. dual exhaust)	n/c

All Firebirds featured this large, contoured type of frameless, all-glass hatch. An electric hatch release was available and was standard on the S/E. This black Trans Am was seen at Elkhart Lake. The bird decal on the roof sail panels showed up better in the black-and-gold color scheme.

UPC Code	Description	Retail Price ($)
LU5	Dual EFI engine package (avail. on Trans Am only; incl. 5.0-liter dual EFI 305-ci Crossfire-Injection V-8 engine, 215/65R15 blackwall steel-belted radial tires, special handling package, limited-slip differential, sport hood, front and rear power disc brakes, and dual exhaust system)	
	w/o Y84	899.00
	w/Y84	n/c
MM4	Four-speed manual trans. (na on base Firebird w/LG4)	n/c
MX1	Automatic trans.	
	Firebird and Firebird S/E	396.00
	Trans Am	72.00

Approximately 2,000 of the 1982 Trans Ams were fitted with a new Recaro interior option. This was available only on black cars with gold accents and included Parella cloth Recaro front bucket seats, the hatch roof, WS6 or WS7 Special Performance packages, and one of two available 5.0-liter V-8s. The option price was $2,486 with the standard four-barrel V-8 with four-speed manual gearbox. It cost $2,968 when ordered for cars with the Crossfire fuel injected V-8 and automatic transmission. Special black exterior door handle inserts carried the Recaro name in gold letters for identification.

UPC Code	Description	Retail Price ($)
C60	Custom A/C (req. w/A01)	675.00
G80	Limited-slip differential axle	
	w/o J65	76.45
	w/J65	45
UA1	Heavy-duty battery	25.00
AK1	Front, rear and front shoulder Custom seat belts (color-keyed)	
	Firebird and Trans Am	
	w/B20	n/c
	w/o B20	26.00

UPC Code	Description	Retail Price ($)
	Firebird S/E	n/c
D42	Cargo security screen	64.00
UE8	Digital quartz clock (avail. w/U63 only; incl. w/all other radios)	60.00
K35	Cruise control w/resume feature	
	w/MX1	155.00
	w/o MM4	165.00
A90	Remote control deck lid release	
	Firebird, Trans Am	32.00
	Firebird S/E	n/c

The S/E's finish was described as a tone-on-tone color concept. Dual sport mirrors in body color were standard. Also standard were 195/75R14 fiberglass-belted radial tires.

One thing identifying the S/E was the lack of a rear decklid spoiler. One could be added for $69 extra.

Note the body-color door handle inserts and lockable fuel filler door, two other S/E features.

UPC Code	Description	Retail Price ($)
C49	Electric rear window defogger	125.00
VJ9	Emission requirements for California	65.00
K05	Engine block heater (avail. w/LQ9 only)	18.00
B57	Custom exterior group (avail. in base Firebird only; Incl. LH remote sport mirrors, roof drip molding, belt reveal molding)	
	w/o CC1	134.00
	w/CC1 (roof drip and belt reveal moldings not incl.)	73.00
N09	Locking fuel filler door	
	Firebird and Trans Am	11.00
	Firebird S/E	n/c
U21	Rally and instrument panel tachometer gauges (incl. trip odometer)	
	Firebird	149.00
	Trans Am and Firebird S/E	n/c
K81	63-amp heavy-duty generator (w/LQ9 and LC1 only)	
	w/o C49	51.00
	w/C49	n/c
K73	70-amp heavy-duty generator (avail. w/LG4; w/o C60 only)	
	w/o C49	51.00
	w/C49	n/c
K99	85-amp heavy-duty generator (na w/LC1)	
	w/LG4	15.00
	w/o LQ9 or LU5	na
	w/o C49	15.00
	w/C49	n/c
A01	Soft-Ray glass, all windows (req. w/C60)	88.00
CC1	Hatch roof w/removable glass panels (na w/BX5/B80)	
	w/o Y84	790.00
	w/Y84	n/c
BS1	Additional acoustical insulation	
	Firebird and Trans Am	39.00
	Firebird S/E	n/c
C95	Dome reading lamp	22.00
TR9	Lamp group	45.00
B48	Luggage compartment trim	123.00
B34	Carpeted front floor mats	20.00
B35	Carpeted rear floor mats	15.00
D35	Sport OSRV mirrors (LH remote, RH convex manual)	
	Firebird	
	w/o B57	48.00
	w/B57	n/c
	Firebird S/E and Trans Am	n/c

UPC Code	Description	Retail Price ($)
DG7	Sport outside rearview mirrors (LH power, RH convex power)	
	Firebird	
	w/o B57	137.00
	w/B57	89.00
	Firebird S/E Trans Am	89.00
D34	RH visor vanity mirror	7.00
B84	Vinyl bodyside molding	
	Firebird and Trans Am (black)	na
	Firebird S/E (color-keyed)	na
B93	Door edge guards (Firebird; bright)	15.00
B91	Door edge guards (Firebird S/E, Trans Am; black)	15.00
B80	Roof drip moldings (Firebird; n.a. w/CC1; bright; incl. belt reveal molding)	
	w/o B57	61.00
	w/B57	n/c
BX5	Roof drip moldings (Firebird S/E, Trans Am; n.a. w/CC1; black)	29.00
J65	Front and rear power disc brakes (incl. G80; n.a. on base Firebird)	
	w/o WS6	255.00
	w/WS6	n/c
AU3	Power door locks	106.00
AC3	Six-way power driver seat	197.00
A31	Power windows	165.00
V08	Heavy-duty radiator	40.00
U75	Power antenna	
	w/o opt. radios	85.00
	w/opt. radios	55.00
U63	Delco AM radio system	102.00
U69	Delco AM/FM radio system	232.00
U58	Delco AM/FM stereo radio system	317.00
UN3	Delco AM/FM stereo cassette system	411.00
UM6	Delco AM/FM ETR cassette system	606.00
UN9	Radio accommodation package	
	w/opt. radios	n/c
	w/o U75 w/o opt. radios	39.00
	w/U75 w/o opt. radios	9.00
UP8	Dual front and rear speakers (w/U63/U69; monaural radios only; Incl. rear speakers w/extended range)	79.00
Y99	Rally tuned suspension (avail. w/Firebird only; QYA or QYC req. at extra cost)	408.00
Y84	Recaro Trans Am option (avail. w/Trans Am only)	
	w/LU5 (req. w/MX1)	2,968.00
	w/LG4 (req. w/MM4)	2,486.00
AR9	Bucket seats (Firebird, Trans Am)	
	Vinyl Derma trim	n/c

Shown here are three 1982 Firebirds: S/E (top), Trans Am (center), and first-level (bottom). The base car has the optional Rally V wheel covers, which were $144 extra. The lockable gas filler door and the rear wiper were standard only on the S/E.

Otherwise, these cars appear to be in their standard equipment forms. This Trans Am is a pre-production pilot model. Pontiac considered using the name T/A to get out of paying a $5-per-car royalty to the SCCA, but eventually dropped the idea.

UPC Code	Description	Retail Price ($)
	Cloth Pompey trim	28.00
WS6	Special Performance package (na w/base Firebird) Trans Am (incl. Sport hood and black wheel covers; Optional 15x7 in. open-finned turbo-cast aluminum wheels Avail. n/c, n.a. w/Y84)	
	w/o LU5 and Y84	417.00
	w/LU5 or Y84	n/c
	Firebird S/E (incl. P20)	387.00
WS7	Special Performance package (na base Firebird) Trans Am (incl. sport hood and black wheel covers; Opt. 15x7 in. open-finned turbo-cast aluminum wheels Avail. n/c, n.a. w/Y84)	
	w/o LU5, Y84	238.00
	w/LU5, Y84	n/c
	Firebird S/E (incl. P20)	208.00
D80	Rear deck spoiler (na base Firebird)	
	Firebird S/E	69.00
	Trans Am	n/c
N33	Tilt steering wheel	95.00

UPC Code	Description	Retail Price ($)
B20	Luxury trim group (incl. luxury front and rear seats, luxury Doors, custom seat belts, map pocket)	
	Firebird	
	Doeskin vinyl or parella cloth	299.00
	Leather	844.00
	Trans Am (w/o Y84)	
	Doeskin vinyl or parella cloth	299.00
	Leather	844.00
	Trans Am (w/Y84)	
	Parella cloth	n/c
	Firebird S/E	
	Doeskin vinyl or parella cloth	n/c
	Leather	545.00
P06	Wheel trim rings (avail. w/base Firebird only)	37.00
PE5	Rally V wheel covers (avail. w/base Firebird only)	229.00
N90	Cast aluminum wheels (avail. w/base Firebird only; silver, gold)	375.00
N24	15x7-in. finned turbo-cast aluminum wheels (avail. w/WS6, WS7 only)	n/c
P20	Bright aluminum wheel hubcap	

Cut-away view of the 1982 Trans Am shows features such as modified MacPherson strut front suspension, torque arm rear suspension, tilt steering controls and basic drivetrain, chassis layouts, and interior packaging.

UPC Code	Description	Retail Price ($)
	(avail. w/Firebird S/E and Trans Am only; incl. w/WS6, WS7 on Firebird S/E)	n/c
PB4	Wheel locking package (avail. w/base wheel on Firebird S/E, Trans Am; avail. w/N90 on base Firebird)	16.00
C25	Rear window wiper/washer system	
	Firebird, Trans Am	117.00
	Firebird S/E	n/c
CD4	Controlled-cycle windshield wiper system	47.00
QYF	195/75R14 blackwall fiberglass-belted radial tires (Firebird)	n/c
QYG	195/75R14 whitewall fiberglass-belted radial tires (Firebird)	62.00
QXV	195/75R14 blackwall steel-belted radial tires (Firebird)	64.60
QVJ	195/75R14 whitewall steel-belted radial tires (Firebird)	126.60
QXQ	195/75R14 white-letter steel-belted radial tires (Firebird)	148.60
QYA	205/70R14 blackwall steel-belted radial tires (na w/WS6)	
	Firebird w/Y99	123.76
	Firebird S/E, Trans Am	n/c

UPC Code	Description	Retail Price ($)
QYC	205/70R14 white-letter steel-belted radial tires (na w/WS6)	
	Firebird w/Y99	211.76
	Firebird S/E, Trans Am	88.00
QYZ	215/65R15 blackwall steel-belted radial tires (avail. w/WS6 only; Firebird S/E and Trans Am)	n/c
QYH	215/65R15 white-letter steel-belted radial tires (avail. w/WS6 only; Firebird S/E, Trans Am)	92.00

1983

For 1983, the Firebirds had some new engines and transmissions, suspension refinements, slight interior revisions, and minor equipment changes, including 13 new and 5 revised options. But the big news for far-sighted collectors was a pair of new option-created models and a midyear high-output (HO) V-8 engine.

The first of the new cars was a 25th Anniversary Daytona 500 Limited Edition Trans Am, designed to commemorate PMD's 25th season as pace car for the Daytona 500 stock car race.

On its exterior, the car featured special mid-body two-tone paint with white upper finish and

One of the most successful competitors in the 1982 SCCA Trans Am Championship Race series was driver Elliott Forbes-Robinson, piloting the yellow and blue STP Sun of a Gun Pontiac Trans Am prepared by Huffaker Engineering.

Pontiac motorsports activities picked up steam in 1982 at courses like Road America, where this race car was photographed in the paddock area. The tubular chassis frame construction suggests how far-removed from production specifications the Trans Am race cars were.

midnight sand gray metallic lower finish. It included a special aero package consisting of rocker panel extensions, front and rear fascia extensions, air dam, rocker fences, grille pads, and special covers on its 15x7-inch turbo-aero aluminum wheels. There was also a sport hood appliqué, a locking fuel filler door, tinted glass, a power antenna, and special Daytona 500 25th Anniversary graphics.

Inside the pace car were special light sand gray Recaro front bucket seats with leather bolsters and headrests, and medium sand gray pigskin inserts. The sides and backs of the seats were upholstered in Pallex cloth. Matching door panels were light sand gray with Pallex cloth door inserts. On the standard equipment list were red instrument panel lighting, AM/FM ETR stereo cassette sound system with graphic equalizer, air conditioning, cruise control, remote-control deck lid release, electric rear window defogger, added acoustical insulation, lamp group and dome reading light, luggage compartment trim, special Daytona 500 floor mats, power mirrors, window and door locks, a tilt steering wheel, controlled-cycle windshield wipers, and a leather-wrapped steering wheel.

There were three power team combinations, with California limited to one. In federal areas, buyers could choose a 5.0-liter four-barrel V-8 with a five-speed manual transmission or 5.0-liter dual-EFI V-8 with automatic. For California, the sole choice was the 5.0-liter four-barrel; with automatic transmission. Also included was the WS6 Special Performance package, including four-wheel disc brakes and a limited-slip differential.

According to the press release that announced this special model on November 1, 1982, production of 2,500 units was scheduled. The price of this model was in the $18,000 range.

A second special model for 1983 was called the Special Edition Recaro Trans Am. This equipment group was marketed as an option package (Y84), although it was promoted as a separate model in the November 1 press release. The package price was $3,160 with the 5.0-liter four-barrel engine and five-speed manual gearbox. When the buyer ordered the Cross Fire Injection powerplant teamed with its mandatory automatic transmission, the price jumped to $3,610.

On its exterior, the Recaro Trans Am included black upper-body and gold lower-body finishes with gold-letter Recaro tape inserts on the door handles and cloisonné gold-and-black birds on the sail panels. Gold-finished, finned, turbo-cast aluminum wheels were fitted with P215/65R15 black-wall, steel-belted radial tires. There was also a gold sport hood appliqué and a removable hatch panel roof.

The interior consisted of Recaro front bucket seats with adjustable thigh and lumbar support. There were doeskin leather seat inserts, bolsters

For 1983, the S/E model included the addition of a folding split-back seat. However, custom pedal trim and a rear hatch wiper and washer were no longer standard. The rear wiper, rear spoiler and white-letter tires shown were extra-cost items. Note S/E identification on sail panel.

and headrests, and doeskin and Pallex cloth side facings and seatbacks. The doors were trimmed with Pallex cloth inserts, and an AM/FM stereo cassette radio with graphic equalizer was featured. Also standard was the WS6 suspension.

In terms of appearance, the regular Firebird threesome was essentially unchanged for 1983, although new dimensions were given, on specification sheets, for height (50.7 inches) and width (72.6 inches). The first-level coupe price was up to $8,813 with the base four-cylinder engine. It had no changes in exterior appearance, but included a new wide-ratio four-speed manual transmission with an integral rail shifter. There were also new standard Rally wheels with exposed black lug nuts and, inside, an improved dual-retractor seat belt system. No longer listed as standard equipment was a front stabilizer bar.

New for the Firebird S/E was a steeper ($10,736) price tag. It partially reflected the change to a new standard powerplant: a 2.8-liter HO V-6 with larger intake and exhaust valves, low-restriction exhaust valves, low-restriction exhaust system, higher 8.9:1 compression ratio, and special pistons. This engine was rated at 135 net horsepower at 5,400 rpm compared to the 1982 engine's 112 net horsepower. It came

attached to the new, five-speed manual gearbox.

Optional powerplants on the 1983 S/E included the 2.5-liter EFI four-cylinder as a delete-option or the 5.0-liter four-barrel with a transmission choice of five-speed manual or automatic.

Inside the S/E, a new Pallex cloth upholstery combination was standard along with a split folding rear seat and separate adjustable headrests for the reclining buckets. These items were also part of the Custom trim option for first-level and Trans Am models. In addition, buyers could pay extra for Lear Siegler adjustable Custom front bucket seats with luxury cloth or leather upholstery. A redesigned, leather-wrapped Formula steering wheel was optional in Firebirds and standard in S/E and Trans Am models.

Different standard suspension setups were used for S/E V-6s and V-8s. The V-6 suspension featured P205/70R14 steel-belted tires on 14x7-inch cast-aluminum wheels, recirculating ball steering gear with a 14:1 ratio, 30-mm front stabilizer bar, 12-mm rear stabilizer bar, 64-Newton millimeters (N/mm) front springs, and 18/25-N/mm variable-rate rear springs. With V-8 engines, the front springs were upgraded to 70-N/mm specifications.

The 1983 Trans Am had a new five-speed manual transmission that was standard for this model and the Firebird S/E. Base engine was the 5.0-liter, four-barrel V-8 with dual resonators and tail pipes.

With the Trans Am selling for $10,810, the price spread between the S/E and the Trans Am more than doubled, though it still wasn't big. The standard extras for this model were unchanged from 1982, except for power teams and suspension details and the new-design Formula steering wheel. The standard engine was again the 5.0-liter four-barrel V-8, which gained 10 net horsepower. It was linked to the five-speed manual gearbox, which featured a 2.95:1 first gear and a 0.73:1 fifth gear, plus standard 3.23:1 rear axle for equal performance with better fuel mileage. Another improvement was that the fresh-air induction system, featuring an asymmetric sport hood (off-center power blister), was standard in all Trans Ams.

Trans Am suspension choices started with the same setup used in V-8 powered S/Es. This was called the Level II V-8 suspension. Level III was the WS6 Special Performance option, which was

available for all Firebirds with high-output V-6 or V-8 engines. The option included P215/65R15 steel-belted radials, 15x7-inch wheels, recirculating ball steering gear with 12.7:1 ratio, 32 mm front stabilizer bar, 21-mm (V-8) or 18-mm (HO V-6) rear stabilizer bar, 96-N/mm front springs, 23.5-N/mm rear springs, four-wheel power disc brakes, limited-slip rear axle, and Stowaway spare tire.

For all Firebirds in 1983, the optional automatic transmission became the new four-speed overdrive model featuring a lockup torque converter for better high-speed economy and greater efficiency. It was first used with V-8s only and later made available with fours and V-6s.

On June 10, 1983, Pontiac announced the release of a HO V-8 engine option for the Trans Am. This powerplant was a third 5.0-liter rated for 190 net horsepower at 4,800 rpm and offering a peak torque rating of 240 foot-pounds at 3,200 rpm. Surprisingly, it used a four-barrel carburetor, instead of the TBI system. This L69 engine and five-speed gearbox WS6 package retailed for $913.

Improved output of this engine, over the 5.0-liter TBI V-8, was attributed to a higher-lift, higher-rpm camshaft; a new, freer-breathing exhaust system; electronic spark control; 9.5:1-compression-ratio pistons and cylinder heads; a

Option DX1 was this "venetian blind" Trans Am hood appliqué on the raised portion of the standard air induction hood. This option was a no-charge feature for Special Edition Recaro Trans Ams and $38 extra on others. Basketball's Moses Malone, poking through the open driver's-side hatch roof, was awarded this 1983 Trans Am for being the National Basketball Association's Most Valuable Player.

These adjustable Lear Siegler front bucket seats are shown here in Pallex cloth trim; they also came in leather trim at a higher cost. A gauge cluster featuring full instrumentation, tachometer, and trip odometer (but not a clock) was standard with Trans Ams as shown here.

fresh-air hood-induction system; and the use of an electric motor-driven fan in place of an engine-drive fan. The latter change, alone, was said to be good for 10 extra horsepower. Road & Track tested such a car at 16 seconds (85 miles per hour) in the quarter mile. It did 0–60 miles per hour in just 7.9 seconds.

Other drivetrain components used with the HO V-8 included the five-speed manual gearbox and a 3.73:1 rear axle ratio. Also standard with the HO engine option was the WS6 Special Performance package, including four-wheel disc brakes, P215/60R15 Goodyear Eagle GT tires, and 15x7-inch turbo-aero aluminum or turbo-finned aluminum wheels.

Specific suspension tuning was another feature of the HO Trans Am. This setup incorporated larger 32-mm front stabilizer bar, higher rate bushings in the control arms and track bar, specific shock absorber calibration, added body-to-frame structural reinforcements, and faster (12.7:1) power steering gear with higher-rate torsion bar. A stiffer torque arm, connecting the transmission to the rear axle, was added to the rear suspension and some lighting holes in the frame were eliminated to give extra strength and stiffness.

Other new options for 1983 included 66-amp and 78-amp heavy-duty alternators, electric outside rearview sport mirrors, sport stripes for first-level Firebirds, a louvered rear window sun shield, and redesigned dual rear speakers. There were many other modifications to interior and sound system option.

The cars continued to be coded using the 17-character VIN system with the format: 1G2A (Model) ( ) ( ) D ( ) 200001 and up. The 1 meant General Motors, the G indicated gas-powered, the 2 meant PMD, the A indicated type of restraint system (passive restraints).

A new Trans Am engine option was a 5.0-liter HO tuned for 190 horsepower at 4,800 rpm. With five-speed and 3.73:1 rear axle, the car went to 60 miles per hour in seven seconds. Note the turbo finned cast-aluminum wheels, which were $325 extra for Firebirds and a no-cost option for S/Es and Trans Ams.

Model numbers were the same as used in 1982 and were followed by an engine code in the first blank bracket. Next came the check digit, then a D for 1983, followed by the assembly plant code and the six-digit sequential production number, which started with a 2. Engine codes were as follows:

Engine	Code
2.5-liter four	F, R, 2
2.8-liter V-6	7, X, 1
2.8-liter HO V-6	Z, L
5.0-liter V-8 4V	G, H, S

After strong initial sales in 1982, Firebird assemblies tapered off to 74,884 units for the 1983 model year. This dropped the Firebird from the eighth-ranked U.S. compact to the ninth. Of the total produced, 39,928 (53.32 percent) had V-8s, 27,333 (36.5 percent) had V-6s, and only 7,623 (10.18 percent) had four-cylinder engines. Automatic transmissions were used in 60,956 (81.4 percent) of the 1983 models and five-speed manuals in 13,928 (18.6 percent).

The performance and handling characteristics of the 1983 Firebirds were very similar to those of 1982 models. However, the new engine and transmission combinations seemed to help fuel economy quite a bit.

In his article in the October 1983 issue of *High-Performance Pontiac,* writer Roger Hunting-ton included calculated performance data for a 1983 Trans Am with the optional 190 net horsepower 5.0-liter HO V-8. Huntington determined that the 3,300-pound car would be good for a top speed of 132 miles per hour and suggested 0–60-mile-per-hour time of nine seconds. His calculations showed 16.9 seconds for a quarter-mile run with an 85-mile-per-hour terminal speed.

Editors of *Consumer Guide* gave a two-week-long road test to the Trans Am with the carbureted engine and automatic transmission, recording a slower 0–60 time of 10.8 seconds. Their report said the car offered "instant throttle response" and "plenty of low end torque," and noted it was easy to drive in traffic but had safe passing reserves for highway use.

Consumer Guide also praised the car's road manners. Among favored features were a well-tuned suspension, accurate and responsive steering, agile handling, and the thick anti-roll bar at the rear, which neutralized the Firebird's front weight bias. The brakes were found to be "impressive despite spongy pedal action."

Fuel economy, which had been as low as 14.5 miles per gallon in normal driving for 1982 *(Road & Track),* was now up to as much as 24 miles per gallon city and 39 miles per gallon highway in the 1983 four-cylinder models with five-speed transmission. Pontiac published the following 1983 Firebird projected EPA fuel economy estimates:

Engine	Transmission	Certification	City	Highway	Combined
2.5-liter L-4 EFI	four-speed manual	federal, California	24	34	28
	five-speed manual*	federal htr.	24	39	30
		federal, California	23	37	28
	three-speed automatic	federal, high-altitude, California	23	35	27
2.8-liter V-6	five-speed manual	federal, California	20	34	25
	three-speed automatic	federal, California	20	36	25
2.8-liter HO V-6	five-speed manual	federal, California	20	34	25
	four-speed automatic	federal, California	20	36	25
5.0-liter V-8	five-speed manual	federal, California	15	24	18
	four-speed automatic	federal, California	17	28	21
5.0-liter dual EFI	four-speed automatic	federal, California	16	27	20

*Fuel economy leader

1983 Firebird Options and Accessories

UPC Code	Description	Retail Price ($)
LQ9	2.5-liter EFI 4-cyl.	
	Firebird	n/c
	Firebird S/E (na w/WS6) (credit)	($300.00)
LC1	2.8-liter V-6 2 bbl (Firebird)	150.00
LL1	2.8-liter HO V-6 2 bbl	
	Req. w/MX0 w/VJ9; Firebird S/E)	n/c
LG4	5.0-liter V-8 4 bbl (na w/MM4 or MX1)	
	Firebird	
	w/MX0	350.00
	w/MM5 (incl. dual exhaust)	375.00
	Firebird S/E	
	w/MX0 w/o WS6	50.00
	w/MM5 or WS6 (incl. dual exhaust)	75.00
	Trans Am (incl. dual exhaust)	n/c
LU5	5.0-liter crossfire EFI V-8 package	
	(avail. on Trans Am w/MX0 only)	
	w/o Y84	858.00
	w/Y84	n/c
MM4	Four-speed manual transmission	
	(avail. w/LQ9 only; n.a. Firebird S/E)	n/c
MM5	Five-speed manual transmission (na w/LU5)	
	Firebird	125.00
	Trans Am	n/c
	Firebird S/E	n/c
MX1	Three-speed automatic transmission	
	Firebird (avail. w/LQ9 or LC1 only)	425.00
	Firebird S/E (avail. w/LQ9 only)	195.00
MX0	Four-speed automatic transmission	
	Firebird (avail. w/LG4 only)	525.00
	Firebird S/E (avail. w/LL1 or LG4 only)	295.00
	Trans Am	295.00
C60	Air conditioning (req. w/A01)	725.00
G80	Limited-slip differential axle	
	w/o J65	95.43
	w/J65	.43
UA1	Heavy-duty battery	25.00
AK1	Color-keyed seat belts	
	Firebird, Trans Am	
	w/o B20	26.00
	w/B20	n/c
	Firebird S/E	n/c
D42	Cargo security screen	64.00
V08	Heavy-duty cooling system	
	w/o C60	70.00
	w/C60	40.00
K35	Cruise control w/resume feature	170.00
A90	Remote-control deck lid release	
	Firebird, Trans Am	40.00
	Firebird S/E	n/c

UPC Code	Description	Retail Price ($)
C49	Electric rear window defogger	135.00
VJ0	Emission requirements for California	75.00
K05	Engine block heater	18.00
B57	Custom exterior group (avail. w/base Firebird only; Incl. LH remote and RH manual sport mirrors, door handle Ornamentation, roof drip molding, belt reveal molding)	
	w/o CC1	112.00
	w/CC1	
	(roof drip, belt reveal moldings not incl.)	51.00
D53	Hood bird decal (Trans Am only)	n/c
DX1	Hood appliqué (Trans Am only)	n/c
	w/o Y84	38.00
	w/Y84	n/c
N09	Locking fuel filler door	
	Firebird, Trans Am	11.00
	Firebird S/E	n/c

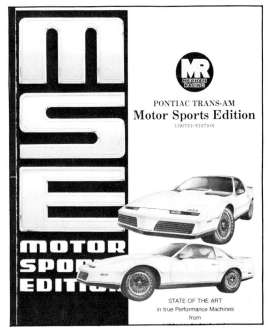

Mecham Racing, Inc., of Tacoma, Washington, produced a street version of its well-known SCCA race machines. To these well-optioned Trans Ams, Mecham added Koni shocks and struts, P255/55R13 HPR Firestone tires, rear droop limiter straps, special alloy wheels, a leather-wrapped "Duesenberg" steering wheel, custom identification badges, Motorola sound equipment, and much more.

UPC Code	Description	Retail Price ($)
U21	Rally gauge cluster w/tachometer & trip odometer	
	Firebird	150.00
	Trans Am, Firebird S/E	n/c
K81	66-amp heavy-duty generator (avail. w/LC1 and LL1 w/o C60 only)	
	w/C49	n/c
	w/o C49	51.00
K64	78-amp heavy-duty generator (avail. w/LQ9 and LG4 w/o C60 only)	
	w/LG4, C49	n/c
	w/LQ9, C49	25.00
	w/o C49	51.00
K99	85-amp heavy-duty generator (avail. w/LG4, LU5, C60 only)	
	w/LU5, C49	n/c
	All other applications	25.00
A01	Soft-Ray glass, all windows (req. w/C60)	105.00
Y99	Rally tuned suspension handling pkg. (avail. base Firebird only; req. w/QYA, QYC; n.a. w/N90, N91)	50.00
CC1	Hatch roof w/removable glass panels (na w/BX5/B80)	
	w/o Y84	825.00
	w/Y84	n/c
BS1	Additional acoustical insulation	
	Firebird, Trans Am	40.00

UPC Code	Description	Retail Price ($)
	Firebird S/E	n/c
C95	Dome reading lamp	23.00
TR9	Lamp group	34.00
B48	Luggage compartment trim	123.00
B34	Carpeted front floor mats	20.00
B35	Carpeted rear floor mats	15.00
D35	Sport OSRV mirrors (LH remote, RH manual)	
	Firebird	
	w/o B57	51.00
	w/B57	n/c
	Firebird S/E, Trans Am	n/c
DG7	Sport electric OSRV mirrors (LH & RH)	
	Firebird	
	w/o B57	140.00
	w/B57	89.00
D34	RH visor vanity mirror	7.00
B84	Vinyl bodyside moldings	
	Firebird, Trans Am (black)	55.00
	Firebird S/E (color-keyed)	n/c
B93	Door edge guards (Firebird; bright)	15.00
B91	Door edge guards (Firebird S/E, Trans Am; black)	15.00
B80	Roof drip moldings (Firebird; n.a. w/CC1; bright; incl. belt reveal moldings)	
	w/o B57	61.00
	w/B57	n/c
BX5	Roof drip moldings (Firebird S/E, Trans Am;	

Introduced in the middle of the 1983 model run was the Silver Anniversary Daytona 500 package for the Trans Am. This model-option cost around $18,000. A total of 2,500 were scheduled for assembly. If you want a collectible, factory-built 1983 Trans Am, this is the one to look for.

UPC Code	Description	Retail Price ($)
	n.a. w/CC1; black)	29.00
J65	Front & rear power disc brakes (incl. G80; n.a. base Firebird)	
	w/o WS6	274.00
	w/WS6	n/c
AU3	Power door locks	120.00
AC3	Six-way power driver seat (na w/Y84)	210.00
A31	Power windows	180.00
U63	AM radio system	112.00
UL6	AM radio system w/clock	151.00
UU9	AM/FM ETR stereo radio system	248.00
UL1	AM/FM ETR stereo radio system w/clock	287.00
UU7	AM/FM ETR stereo radio system w/cassette clock	387.00
UU6	AM/FM ETR stereo radio w/cassette, seek-and-scan, Graphic equalizer & clock	
	w/o Y84	590.00
	w/Y84	n/c
UN9	Radio accommodation package	
	w/opt. radios	n/c
	w/o U75 w/o opt. radios	39.00
	w/U75 w/o opt. radios	9.00
UP8	Dual rear speakers (avail. w/UL6 only)	40.00
U75	Power antenna	
	w/o opt. radios	90.00
	w/opt. radios	60.00
Y84	Special Edition Recaro Trans Am opt. (avail. w/Trans Am only)	
	w/LU5 (req. w/MX0)	3,610.00
	w/LG4 (req. w/MM5)	3,160.00
WS6	Special Performance package (na w/base Firebird) Trans Am	
	w/o LU5, Y84	408.00
	w/LU5, Y84	n/c
	Firebird S/E	408.00
AR9	Bucket seats (Firebird, Trans Am)	
	Oxen vinyl trim	n/c
	Pompey cloth trim	30.00
D80	Rear deck spoiler	
	Firebird, Firebird S/E	70.00
	Trans Am	n/c
NP5	Leather-wrapped Formula steering wheel	40.00
N33	Tilt steering wheel	105.00
D98	Vinyl sport stripes (Firebird only)	75.00
DE1	Louvered rear window sunshield, hinge-mounted (na w/C25)	210.00
B20	Luxury trim group (incl. front and rear luxury seats, luxury doors, Color-keyed seat belts, split folding rear seat, map pocket, carpeted cowl kick panel)	
	Luxury reclining bucket seats (Pallex cloth trim)	

UPC Code	Description	Retail Price ($)
	Firebird	349.00
	Trans Am	349.00
	Firebird S/E	n/c
	Lear Siegler bucket seats, adjustable (Pallex cloth trim)	
	Firebird	749.00
	Trans Am	749.00
	Firebird S/E	400.00
	Lear Siegler bucket seats, adjustable (leather trim)	
	Firebird	1,294.00
	Trans Am	1,294.00
	Firebird S/E	945.00
P06	Wheel trim rings (avail. w/base Firebird only)	38.00
PE5	Rally V wheel covers (avail. w/base Firebird only)	95.00
N91	Wire wheel covers w/locking package (avail. w/base Firebird only; n.a. w/Y99)	185.00
N90	Cast aluminum wheels (avail. w/base Firebird only; n.a. w/Y99)	225.00
N24	Turbo-finned cast aluminum wheels	
	Firebird (avail. only w/Y99)	325.00
	Trans Am, Firebird S/E	n/c
PB4	Wheel locking package (avail. Trans Am and Firebird S/E; avail. w/base Firebird w/N24 or N90 only)	16.00
C25	Rear window wiper/washer system	120.00
CD4	controlled-cycle windshield wiper system	49.00
QYF	195/75R14 blackwall fiberglass-belted radial tires (Firebird)	n/c
QYG	195/75R14 whitewall fiberglass-belted radial tires (Firebird)	62.00
QXV	195/75R14 blackwall steel-belted radial tires (Firebird)	126.48
QXQ	195/75R14 white-letter steel-belted radial tires (Firebird)	148.48
QYA	205/70R14 blackwall steel-belted radial tires (na w/WS6)	
	Firebird (w/Y99 only)	122.68
	Firebird S/E, Trans Am	n/c
QYC	205/70R14 white-letter steel-belted radial tires (na w/WS6)	
	Firebird (w/Y99 only)	210.68
	Firebird S/E, Trans Am	88.00
QYZ	215/65R15 blackwall steel-belted radial tires (avail. w/WS6 only; Firebird S/E, Trans Am)	1.40
QYH	215/65R15 white-letter steel-belted radial tires (avail. w/WS6 only; Firebird S/E, Trans Am)	93.40

1984

With a modest price increase, the 1984 first-level Firebird climbed to $8,753, the S/E sold for $11,053, and the standard equipped Trans Am had a base retail price of $11,103. Seasonal changes followed the pattern of refinements to paint, upholstery, suspension, and engines. For future collectors, the Recaro option-created model returned and the 15th Anniversary Trans Am was released early in the year.

New exterior equipment standard on the first-level Firebird included a black-finished-mast antenna, fourth-generation all-weather radial tires, and tinted rear hatch glass. The interior features list included dual horns and an upshift indicator light with manual transmissions. Two-tone paint and a rear decklid spoiler were newly-available options.

The base 2.5-liter four-cylinder engine was back with new swirl-port cylinder heads and a 9.0:1 compression ratio. It was rated 92 net horsepower at 4,400 rpm. Four-speed manual transmission was standard everywhere, and the five-speed, with a new hydraulic clutch, was available outside California. Power options included the regular 2.8-liter V-6 and the 5.0-liter V-8.

There were no changes in the exterior features that distinguished 1984's version of the "subtle and sophisticated" Firebird S/E. Inside was a new color-keyed, leather-wrapped Formula steering wheel and redesigned bucket seat headrests. Also included was a full set of gauges including tachometer and trip odometer.

The Level II Rally tuned suspension with P205/70R15 all-weather tires, 14x17-inch turbo aluminum wheels, and larger front and rear stabilizer bars was basic too. In 1984, the Level II suspension had polyurethane front and rear stabilizer grommets added to reduce body roll. In addition, all suspensions used higher-rate bar bushings for improved lateral restraint of the rear axle.

Base powerplant in the S/E was again a 2.8-liter HO V-6. It featured two-barrel carburetion and an 8.9:1 compression ratio, and it had 15 more horsepower (125 at 5,400 rpm) than the previous season. A five-speed manual transmission was standard in federal areas. Since four-speed manual transmissions were limited to first-level Firebirds, an automatic transmission was a mandatory options for S/Es (except those with "delete" engine options) sold in California. The S/E could be ordered with the 2.5-liter four as a delete-option. Also available at extra cost was the regular 5.0-liter V-8.

A slick-looking Aero package consisting of front and rear fascia extensions, rocker panel

New for the 1984 Trans Am (only) was the W62 Aero package, a no-cost option for Recaro models. This car does not have Recaro seats. You can tell by the design of the headrests seen through the window. In this case, the Aero package was $199 extra.

extensions, and fade-away lower body-side accent stripes was a new $200 option for 1984 Trans Ams. Exterior equipment features for the base Trans Am no longer included front wheel-opening flares (rear flares were still used, however), and aero turbo-cast aluminum wheels replaced the 1983's black-finished-with-center-cap design.

Big news for Trans Ams was the continuation of the high-output V-8 option, including the WS6 suspension package and the five-speed manual gearbox (certified for both federal and California use). The 1984 Trans Am's base powerplant was the regular 5.0-liter V-8 with a 8.5:1 compression and a 150 net horsepower rating. The four-speed automatic was optional with both Trans Am V-8s and could be ordered with all 1984 Firebird engines in specific models.

The Recaro option consisted of Aero package; leather-wrapped Formula steering wheel, shift knob, and parking brake handle; leather-trimmed Recaro luxury front bucket seats; luxury trim; split folding rear seats; luxury door panels; instrument panel map pocket; color-keyed seat belts; 15x7-inch gold-finished high-tech aero wheels; and gold hood appliqué. The Recaro option was available for the Trans Am only, at a price of $1,621 extra.

In its October 1984 issue, *High-Performance Pontiac* highlighted the special, limited-edition 15th Anniversary Trans Am and compared its appearance to the legendary "The Judge" GTO. That is to say, the magazine felt the exterior of the car was a little over-done.

Finished in white with medium blue trimmings, the car included white aero skirting and grille slot covers, white 16-inch aluminum high-tech wheels, and a white rear deck spoiler. The interior had white Recaro bucket seats with blue inserts lettered with the Trans Am name. The Formula steering wheel was wrapped in white with a blue shield-shaped badge that read "Trans Am: 15" in the center of its hub. The same badge appeared on the sail panels, while the Trans Am name, in blue, decorated the lower sides of the body. A hatch roof with dark-colored glass panels set the roof off against the white body, and multiple blue pinstripes traced the lower body perimeter entirely around the car. On the hood was a blue "venetian blind" decal with a white-outlined bird and 5.0-liter HO lettering.

The engine was the 190 net horsepower V-8 linked to a five-speed manual transmission. Also included were the WS6 suspension and special P245/50R16 Goodyear Eagle GT tires.

Pontiac scheduled production of 1,500 of the 15th Anniversary models. Their price tag was $3,499 over a normally equipped WS6 Trans Am. That meant a total window sticker of about $17,500.

High-Performance Pontiac found the car's suspension and tires too stiff and harsh for what it called "real world driving." The engine ran hot and stalled in city traffic conditions. Hard cornering brought on fuel starvation. The special

One way to spot a 1984 Trans Am is to look for the model name on the right-hand corner of the rear bumper panel. Black, full-width taillights were standard again as was a rear decklid spoiler.

Recaro seats were found to be uncomfortable, and body flex caused an undue amount of noise. Gearshift action was loose and sloppy, wheel hop was a problem, and performance was unspectacular. The 0–60-mile-per-hour runs took 7.8 seconds, while the quarter-mile acceleration tests resulted in a top figure of 16.2 seconds with a terminal speed of 86 miles per hour.

Editor Cliff Gromer said, "What we would do if we bought one of these cars would be to sock it away in a garage someplace . . . then 10 years down the road, we'd take a peek at Jerry Heasley's "Collecting" column in this magazine and see what the 15th Anniversary cars are doing on the auction blocks. Chances are we'd turn a fat profit."

As for the performance of other 1984 Firebirds, several magazines and other sources reported varied test run figures. Pontiac claimed that the Trans Am equipped with the 5.0-liter HO V-8 offered an estimated 0–60 capability of around seven seconds. *Road & Track's* figures, in the magazine's *Sports & GT Cars Guide,* gave a 0–60 actual figure of 7.9 seconds, a top speed of 124 miles per hour, and a quarter-mile time of 16.1 seconds. It was also found that the car was fairly noisy and got only 13 miles per gallon in normal driving use.

Car and Driver did not test the 1984 model, but put the 5.0-liter HO through its paces in

April 1985. This car had the same horsepower rating and five-speed manual transmission. Test results were 7.6 seconds for 0–60 miles per hour and 15.6 seconds for the quarter-mile at 87 miles per hour. The only big difference noted was a top speed of 135 miles per hour and 14 miles per gallon "observed" fuel economy rating. (EPA estimates were 15 miles per gallon city and 24 miles per gallon highway.)

After a bad season in 1983, the third-generation Firebird bounced back with much better sales in 1984. Model year production hit 128,304 units. Of these, available equipment breakouts show 121,119 cars with air conditioning, 102,002 with automatic transmission, only 1,925 with four-speed manual transmission, and 24,378 with five-speed manual gearbox.

1984 Firebird Options and Accessories

UPC Code	Description	Retail Price ($)
LQ9	2.5-liter EFI 4-cyl.	
	Firebird	n/c
	Firebird S/E	
	(na w/WS6, WS7, WY6, WY5) (credit)	(350.00)
LC1	2.8-liter V-6 w bbl (na w/MM4; Firebird)	250.00
LL1	2.8-liter HO V-6 w bbl	
	(req. w/MX0 w/VJ9; Firebird S/E)	n/c
LG4	5.0-liter V-8 4 bbl (na w/MM4;	
	req. w/WS6, WS7, WY6 or WY5 w/MM5)	
	Firebird	550.00
	Firebird S/E	200.00
	Trans Am	n/c
L69	5.0-liter HO V-8 4 bbl (avail. w/Trans Am	
	Only; req. w/WS6, WY6, req. w/Y84)	530.00
MM4	Four-speed manual transmission	
	(avail w/LQ9 only; n.a. w/Firebird S/E)	n/c
MM5	Five-speed manual transmission	
	Firebird	125.00
	Trans Am	n/c
	Firebird S/E	n/c
MX0	Four-speed automatic transmission	
	Firebird	525.00
	Trans Am	295.00
	Firebird S/E	295.00
C60	Air conditioning (req. w/A01)	730.00
G80	Limited-slip differential axle	
	w/o WS6/WS7	95.00
	w/WS6/WS7	n/c
UA1	Heavy-duty battery	26.00
AK1	Color-keyed seat belts	
	Firebird, Trans Am	

You're looking at the nerve center of the ultimate Firebird: the Trans Am Recaro Special Edition. Standard interior features include Recaro front buckets with leather in the seating areas and a Delco-GM AM/FM ETR stereo sound system.

The ultimate Firebird was the Y84 Special Edition Recaro Trans Am. This option package cost $1,621.

UPC Code	Description	Retail Price ($)
	w/o B20	26.00
	w/B20	n/c
	Firebird S/E	n/c
D42	Cargo security screen	69.00
V08	Heavy-duty cooling system (na w/L69)	
	w/o C60	70.00
	w/C60	40.00
K34	Cruise control w/resume, accelerate features	175.00
A90	Remote control deck lid release	
	Firebird, Trans Am	40.00
	Firebird S/E	n/c
C49	Elec. rear window defogger	140.00
VJ9	Emission requirements for California	99.00
K05	Engine block heater	18.00
W62	Aero pkg. (Trans Am only)	
	w/o Y84	199.00
	w/Y84	n/c
W51	Black appearance group (base Firebird only; n.a. w/B57)	
	w/o CC1 (incl. black roof drip molding)	152.00
	w/CC1 (black roof drip molding not incl.)	123.00
B57	Custom exterior group (base Firebird only; n.a. w/W51)	
	w/o CC1 (Incl. roof drip & belt reveal molding)	112.00
	w/CC1 (roof drip & belt reveal molding not incl.)	51.00
DX1	Hood appliqué (avail. w/Trans Am only)	
	w/o Y84	38.00
	w/Y84	n/c
N09	Locking fuel filter door	
	Firebird, Trans Am	11.00
	Firebird S/E	n/c
U21	Rally gauge cluster w/tach. Trip odometer Firebird	150.00

In July 1983, Pontiac Motor Division began releasing factory SD parts designed for use with its four-cylinder Iron Duke engine. They turned the little 165-ci engine into a 272-horsepower screamer with a 7,600 rpm power peak! The SD four could move the race-prepped Firebirds to a top speed of 150 miles per hour and through the quarter-mile in 13.2 seconds. At the 1984 Chicago Automobile Show, the company displayed this SD Trans Am with plexiglass hood revealing the hot little powerplant. It was built to IMSA (International Motor Sports Association) Kelly Girl series championship racing specifications.

UPC Code	Description	Retail Price ($)
	Trans Am, Firebird S/E	n/c
K81	66 amp heavy-duty generator (avail. w/LC1, LL1 w/o C60 only)	
	w/C49	n/c
	w/o C49	51.00
K64	78 amp heavy-duty generator (avail. w/LQ9, LG4 w/o C60 only)	
	w/LG4, C49	n/c
	w/LG4 w/o C49	51.00
	w/LQ9, C49	25.00
	w/LQ9 w/o C49	51.00
K22	94-amp heavy-duty generator (avail. w/LQ9/LG4 w/C60 only, L69)	
	w/L69	n/c
	w/LQ9/LG4, C60	25.00
A01	Soft-Ray glass, all windows (req. w/C60)	110.00
CC1	Locking hatch roof w/removable glass panels (na w/BX5/B80)	850.00
BS1	Additional acoustical insulation	
	Firebird, Trans Am	40.00
	Firebird S/E	n/c
C95	Dome reading lamp	23.00
TR9	Lamp group	34.00
B48	Luggage compartment trim	123.00
B34	Carpeted front floor mats	20.00
B35	Carpeted rear floor mats	15.00
D35	Sport OSRV mirrors (LH remote, RH manual) Firebird	
	w/o B57 or W51	53.00
	w/B57 or W51	n/c
	Firebird S/E, Trans Am	n/c

UPC Code	Description	Retail Price ($)
DG7	Sport electric OSRV mirrors (LH and RH) Firebird	
	w/o B57 or W51	139.00
	w/B57 or W51	91.00
	Firebird S/E, Trans Am	91.00
D34	RH visor vanity mirror	7.00
B84	Vinyl bodyside moldings (na w/D84)	
	Firebird, Trans Am (black)	55.00
	Firebird S/E (color-keyed)	n/c
B93	Door edge guards (Firebird w/o W51; bright)	15.00
B91	Door edge guards	
	Firebird w/W51 (black)	15.00
	Firebird S/E, Trans Am (black)	15.00
B80	Roof drip moldings (na w/CC1; Firebird)	
	w/o B57 or W51 (bright; incl. belt reveal molding)	61.00
	w/B57 or W51	n/c
BX5	Roof drip moldings (na w/CC1; Firebird S/E, Trans Am; black)	29.00
D84	Two-tone paint (base Firebird only)	205.00
J65	Front & rear power disc brakes (na w/base Firebird or w/LC1 w/MM5)	
	w/o WS6/WY6	179.00
	w/WS6/WY6	n/c
AU3	Power door locks	125.00
AC3	Six-way power driver seat	215.00

This special 1984 Firebird was announced as a factory model-option in the fall of 1983. Known as the 15th Anniversary Trans Am, it came only with the hatch roof and many other options as well as a special exterior/interior treatment.

Mecham Racing also marketed a 1984 version of its Motor Sports Edition Trans Am. The model name decal was found across the door and windshield, and there was identification on the left-hand headlight door. It looks quite similar to the 15th Anniversary Trans Am, but it was rarer and much more performance oriented.

UPC Code	Description	Retail Price ($)
A31	Power windows	
	Firebird, Trans Am	
	w/o B20	185.00
	w/B20 (incl. door map pockets)	215.00
	Firebird S/E (incl. door map pockets)	215.00
U63	AM radio system	112.00
UL6	AM radio system w/clock	151.00
UU9	AM/FM ETR stereo radio system	248.00
UL1	AM/FM ETR stereo radio system w/clock	287.00
UU7	AM/FM ETR stereo radio system w/cassette clock	387.00
UU6	AM/FM ETR stereo radio w/cassette, seek-and-scan, graphic equalizer, clock	
	w/o Y84	590.00
	w/Y84	n/c
UP8	Dual rear speakers (avail. w/UL6 only)	40.00
UQ7	Subwoofer speaker system (avail. w/UL1, UU6, UU7, UU9 only)	150.00
U75	Power antenna	60.00
AR9	Bucket seats (Firebird and Trans Am)	
	Sierra vinyl trim	n/c
	Pompey cloth trim	30.00
Y84	Special Edition Recaro Trans Am opt.	1,621.00
WS6	Special Performance pkg. (w/J65; n.a. base Firebird; w/G80)	408.00
WS7	Special Performance pkg. (w/o J65; w/G80)	
	Base Firebird	612.00
	Firebird S/E, Trans Am	229.00
WY6	Special Performance pkg. (w/J65; n.a. base Firebird; w/o G80)	313.00
WY5	Special Performance pkg. (w/o J65; w/o G80)	
	Base Firebird	517.00
	Firebird S/E, Trans Am	134.00
D80	Rear deck spoiler	
	Firebird, Firebird S/E	70.00
	Trans Am, W51	n/c
NP5	Leather appointment group	
	Base Firebird, Trans Am (black only)	75.00
	Firebird S/E, Y84 (color-keyed)	n/c
N33	Tilt steering wheel	110.00
D98	Vinyl sport stripes (Firebird only; n.a. w/D84)	75.00
DE1	Louvered rear window sunshield, hinge-mounted (na w/C25)	210.00
Y99	Rally tuned suspension (req. w/QYA or QYC; n.a. w/N91)	
	Firebird	50.00
	Trans Am, Firebird S/E	n/c
B20	Luxury trim group (incl. luxury front & rear seats, luxury doors, color-keyed seat belts, split folding rear seat, map pocket, carpeted cowl kick panel, LH knee pad in Trans Am only)	
	Luxury reclining bucket seats (Pallex cloth trim)	
	Firebird	349.00
	Trans Am	359.00
	Firebird S/E	n/c
	Lear Siegler bucket seats, adjustable Pallex cloth trim)	
	Firebird	749.00
	Trans Am	759.00
	Firebird S/E	400.00
	Lear Siegler bucket seats, adjustable (leather trim)	
	Firebird	1,294.00
	Trans Am	1,304.00
	Firebird S/E	945.00
P06	Wheel trim rings (Firebird)	38.00
P02	Five-port wheel covers (Firebird)	38.00
N91	Wire wheel covers w/locking pkg. (na w/Y99; Firebird)	185.00
N89	Turbo aero cast aluminum wheels (Trans Am, Firebird S/E)	n/c
N24	Turbo finned cast aluminum wheels	
	Firebird w/o WS6, WS7, WY6 or WY5	
	Trans Am, Firebird S/E, base Firebird w/WS6, WS7, WY6 or WY5	n/c
N78	Deep-dish high-tech turbo wheels (req. w/WS6, WS7, WY6 or WY5)	n/c
PB4	Wheel locking package (avail. w/N24, N78, N89)	16.00
C25	Rear window wiper/washer system (na w/DE1)	120.00
CD4	Controlled-cycle windshield wiper system	50.00
QMW	195/75R14 blackwall steel belted radial tires (Firebird)	n/c
QMX	195/75R14 whitewall steel-belted radial tires (Firebird)	62.00
QYA	205/70R14 blackwall steel-belted radial tires (na w/WS6, WS7, WY6 or WY5)	
	Firebird (w/Y99 only)	58.00
	Firebird S/E, Trans Am	n/c
QYC	205/70R14 white-letter steel-belted radial tires (na w/WS6, WS7, WY6 or WY5)	
	Firebird (w/Y99 only)	146.00
	Firebird S/E, Trans Am	88.00
QYZ	215/65R15 blackwall steel-belted radial tires (na w/WS6, WS7, WY6 or WY5 only)	n/c
QYH	215/65R15 white-letter steel-belted radial tires (na w/L69; avail. w/WS6, WS7, WY6 or WY5 only)	92.00

1985

Tuned Port Injection (TPI) fuel injection, improved ride and handling, and a dramatic new look for the Trans Am were highlights of the changes in Firebirds for 1985. The first-level model had its base price increased to $9,263, while the S/E jumped to $11,709 and the standard Trans Am was $11,983.

Rally tuned suspension was made standard on the entry-level Firebird, and the instrument panel and console were redesigned for a softer, more rounded look. The console received a new covering of soft vinyl material and was split in two parts. The pod containing the radio was attached to the instrument panel. New standard features

on the exterior included bird decals on the hood and sail panel, black taillight bezels, and black-finished front fascia pads.

A Delco AM radio was standard equipment, although it could be deleted for credit. Also standard was a heater, and the formerly black-finished instrument panel was now described as "graphite-finished." The base engine was an improved 2.5-liter Tech Four base engine was an improved 2.5-liter Tech IV four-cylinder mated to a five-speed manual transmission. It featured 88 net horsepower at 4,400 rpm and throttle-body injection (TBI). Options included the 2.8-liter V-6 with multi-port fuel injection (MPFI) and 135 net horsepower at 5,100 rpm, or the

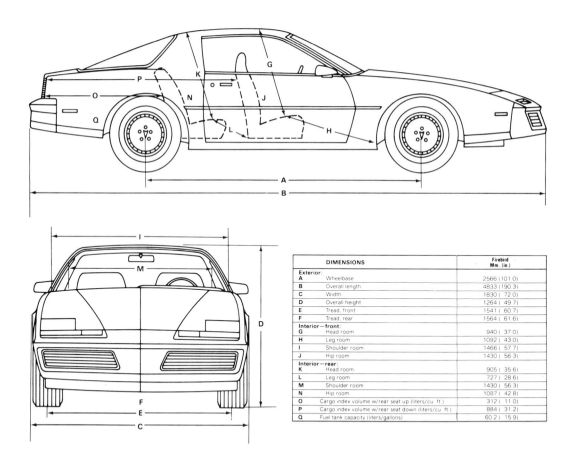

DIMENSIONS		Firebird Mm. (in.)
Exterior:		
A	Wheelbase	2566 (101.0)
B	Overall length	4833 (190.3)
C	Width	1830 (72.0)
D	Overall height	1264 (49.7)
E	Tread, front	1541 (60.7)
F	Tread, rear	1564 (61.6)
Interior—front:		
G	Head room	940 (37.0)
H	Leg room	1092 (43.0)
I	Shoulder room	1466 (57.7)
J	Hip room	1430 (56.3)
Interior—rear:		
K	Head room	905 (35.6)
L	Leg room	727 (28.6)
M	Shoulder room	1430 (56.3)
N	Hip room	1087 (42.8)
O	Cargo index volume w/rear seat up (liters/cu. ft.)	312 (11.0)
P	Cargo index volume w/rear seat down (liters/cu. ft.)	884 (31.2)
Q	Fuel tank capacity (liters/gallons)	60.2 (15.9)

These measurements were essentially the same for all third-generation models, although some specification books show a slight increase in overall length for 1985 due to bumper and skirting varia-

tions. Weights given for the base Firebird V-8s in NADA used car guides were 3,023 pounds in 1982, 3,117 pounds in 1983, 3,157 pounds in 1984, and back down to 3,135 pounds for 1985.

5.0-liter, four-barrel V-8 with 155 net horsepower at 4,200 rpm.

New front and rear fascia designs characterized the 1985 Firebird S/E. Front fascia pads replaced the grille to improve aerodynamics. Black-finished bumper pads, front and rear, added to a more sophisticated appearance. The sail panel now carried bird decals, and the Firebird S/E lettering was on the lower front edge of the doors. New striping accents were featured. Other standard exterior equipment included black-finished door handles and lock cylinders, hood air louvers, neutral-density taillight lenses with a smooth contour look and 14x7-inch diamond-spoke aluminum wheels with a Light Chestnut or Charcoal finish.

A revised interior featured high-contour firm-foam bucket seats upholstered in Pallex cloth, an interior roof console, and Custom trim and door panels. Recaro seats were available as a separate option.

The new 2.8-liter multiple-port fuel injected V-6 was standard equipment and it featured the electric cooling fan. Radiator-mounted fans were designed to use power only when cooling conditions required them. The 5.0 liter carbureted V-8 was an extra-cost option and the four was no longer available for a credit. A modified Y99 suspension setup was also available as an option. All engines came with a choice of five-speed manual or four-speed automatic transmission.

The new Trans Am came with an enlarged version of the 1984 Aero package as standard equipment. It was fully integrated into the front and rear fascias, and front foglights were featured. A new hood had louvers and extractors; it replaced the domed, asymmetrical design. Also used were the neutral density taillights with smooth, aerodynamic surfaces.

The base engine was the 5.0-liter, four-barrel V-8. It had a higher (9.5:1) compression ratio and developed 160 net horsepower at 4,400 rpm. Torque was rated at 250 foot-pounds at 3,200 rpm. Using the same compression ratio was the optional 190 net horsepower engine, which gave 240 foot-pounds of torque at 3,200 rpm. There was also a new TPI (Tuned Port Injection) 5.0-liter option. Also featuring 9.5:1 compression, it

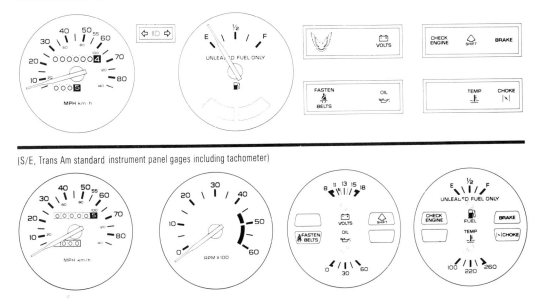

These sketches depict the standard 1985 Firebird instrument panel (top) and standard S/E and Trans Am instrument panel (bottom). The glovebox on the S/E models added a bird emblem at the center.

The Trans Am glovebox had the same emblem, but it was flanked, on the left-hand side, by the model name in red letters.

produced 205 net horsepower at 4,400 rpm and 270 foot-pounds of torque at 3,200 rpm.

The 1985 Trans Am had a very high-tech look enhanced by a better-integrated rear decklid spoiler—finished in black—and its improved aerodynamics. The car sat so low to the ground that a footnote in the sales catalog read "Caution: For use with tire chains see Owner's Manual." Like the other 1985 models, it had bird decals on the sail panels with bright Trans Am letters on the mid-feature-line, directly behind the front wheel openings. Additional standard equipment included P215/65R15 Eagle GT tires on 15x7-inch deep-dish, high-tech wheels (comparable to 1984 WS6 suspension).

In the Firebird option bin, a heavier suspension and 16-inch tires were included with the newly modified 1985 top-level WS6 high-performance setup. P245/50VR16 Goodyear Eagle Gatorback tires, mounted on 16x8-inch high-tech wheels were matched with a larger 34-mm front stabilizer and 25-mm rear stabilizer. Rounding out the package were Delco gas-filled shocks and struts and higher-effort power steering. There was no longer a Recaro model, although both Lear Siegler and Recaro buckets could be ordered as separate options.

There were no straight-out-of-the-box, limited-edition factory Trans Ams for 1985, but there was a genuine Daytona 500 Pace Car. It was Pontiac's fifth straight year of pacing this famous stock car race with a Trans Am.

The pace car was specially equipped with the TPI 5.0-liter V-8 and a four-speed overdrive automatic transmission. It was finished in white with a wide charcoal-colored band and a thin red pinstripe around the entire lower perimeter of the body. Mounted on the roof, behind the glass hatches, was a police-type beacon light. There were also a special hood bird decal, a Pontiac window decal, NASCAR lettering on the rear fender sides, and door lettering reading "Daytona 500: Winston Cup Series; February 17, 1985."

Reaction to the 1985 Firebirds was a mixed bag of criticism and praise. In general, most magazines negatively scored the return of bird decals and extraneous graphic touches. The critics also felt that performance and handling, while improved over the early third-generation models, were not up to the peak levels archived in some other 1985 sports and GT cars. One editor went as far as noting that his Volvo sedan seemed to have more get-up-and-go and just slightly less maximum performance than the Trans Am HO he tested.

The 1985 sales catalog showed the base Firebird in two-tone finish and the front of the S/E on this page. The Firebird name was carried on the front fender. The words Firebird S/E, on a decal, appeared on the door's lower-front corner; S/E emblems on the roof sail panels were no longer used for identification.

As mentioned earlier, *Car and Driver* tried out the 190-horsepower HO in April 1985. Acceleration figures were comparable to those of 1984 models with the same engine, although top speed was 10 miles per hour faster.

In other areas, this magazine was obviously more critical, giving low scores to such things as fit, finish, seating comfort, steering, and suspension. Appearance features like the decal package were criticized for being somewhat dated. And Technical Director Don Sherman advised against ordering the T-top. "It adds weight, compromises structural rigidity, and makes the door difficult to latch," he advised.

Motor Trend (October 1984) gave the new TPI Trans Am a workout. It also disliked the return of the "chicken" hood decal, but made more positive comments about the revised WS6 suspension (mandatory with the 205 net horsepower engine) and the all-new digital electronic instrument panel (*Motor Trend* had seen this option on display, though it was not on its test car).

Performance with the hotter engine (equipped with mandatory four-speed automatic transmission) was in the same performance range as the HO V-8 with the five-speed transmission. The advantage of the 15 extra horsepower was somewhat compromised because GM hadn't yet devised a five-speed manual gearbox that could handle the torque generated by the TPI engine. Associate Editor Tony Assenza registered a top

A new W51 exterior appearance package (left) was available for the 1985 base Firebird. It included black door handles, sport mirrors, rear deck spoiler, and sail panel bird decals. Also available, at extra-cost, was a D84 two-tone paint treatment (right) in 13 different color combinations.

The specific Trans Am Aero package (Code W62) was available for the 1985 Trans Am. This car also has the optional black aero-wing spoiler. New, smooth-contour taillights were covered with a matte-black cross-hatch panel. Pontiac lettering was embossed in the rear panel above the license plate recess. The Trans Am name continued to appear at the right-hand rear corner of the bumper panel. Photo also shows the high-tech turbo aluminum wheels.

0–60-mile-per-hour time of 7.79 seconds and a 16.07-second standing start quarter-mile with an 84.5-mile-per-hour terminal speed. He rated the car as average for its type on the skid pad, but upgraded it to excellent for handling on the road.

While *Car and Driver* had stressed structural weaknesses in the Firebird's body engineering characteristics, *Motor Trend's* Assenza took a different view. He noted improvements that Pontiac had made in body stampings, brackets, and insulation and said his test car was a "quicker, less buzzy, and ultimately more comfortable car to live with" than its 1984 counterpart.

Scott Stevens, of *High-Performance Pontiac*, also tested a TPI Trans Am. Though he had trouble getting sufficient traction off the line (even with the new Gatorback tires), Stevens reported a 15.37 second quarter-mile run at 88 miles per hour. While this was slightly improved over what other magazines achieved, Stevens mentioned the drawback of a 2.29:1 rear axle. With a higher axle

ratio, he guessed that his Trans Am would have been capable of running the quarter-mile in the 14.3 second bracket.

With added weight and added horsepower to most 1985 Firebirds, the EPA fuel economy ratings for various models either stayed about the same or slipped a little bit. The averages released by Pontiac in the fall of 1984 were as follows:

Engine	Horsepower	Transmission	Mileage
2.5-liter EFI 4	88 @ 4400 rpm	5-speed	24/34
		automatic	22/32
2.8-liter MPFI V-6	135 @ 5100 rpm	5-speed	18/27
		automatic	18/26
4.0-liter 4V V-8	155 @ 4200 rpm	4-speed	15/24
		automatic	16/22
5.0-liter HO V-8	190 @ 4800 rpm	5-speed	15/24
5.0-liter TPI V-8	205 @ 4400 rpm	automatic	16/22

For 1985, the Trans Am featured new surface detailing including aero-tuned rocker and quarter panel extensions. There was also a new hood with louvers, air extractors, and built-in fog-lamps. Diamond spoke wheels were a new option priced at $325 for Firebirds and available as a no-cost substitution item for the S/E and Trans Am.

Final model-year production of 1985 Firebirds totaled 95,880 units, a bit less than initially targeted for the year. The most popular 1985 model was the S87 base Firebird coupe, of which 46,644 were built. Surprisingly close behind in the tally was the W87 Trans Am coupe, of which 44,028 left the factory. The rarest model, by a long shot, was the X87 Firebird SE coupe. Only 5,208 of these were manufactured.

1985 Firebird Options and Accessories

UPC Code	Description	Retail Price ($)
LQ9	2.5-liter EFI 4-cyl. (Firebird)	n/c
LB8	2.8-liter EFI V-6	
	Firebird	350.00
	Firebird S/E	n/c
LG4	5.0-liter V-8 4-bbl (MM5 req. w/Y99)	
	Firebird	650.00
	Firebird S/E	300.00
	Trans Am	n/c
L69	5.0-liter HO V-8 4 bbl (avail. w/Trans Am only; n.a. w/MX0; req. w/G80 w/QAC or QDZ)	695.00
LB9	5.0-liter EFI V-8 (avail. w/Trans Am only; Req. w/MX0 w/QAC or QDZ)	695.00
MM5	Five-speed manual transmission (na w/LB9)	n/c

UPC Code	Description	Retail Price ($)
MX0	Four-speed automatic transmission (na w/L69)	425.00
C60	Air conditioning (req. w/A01)	750.00
G80	Limited-slip differential axle	
	w/o WS6	100.00
	w/WS6	n/c
UA1	Heavy-duty battery (na w/LQ9)	26.00
AK1	Color-keyed seat belts	
	Firebird, Trans Am	
	w/o B20	26.00
	w/B20	n/c
	Firebird S/E	n/c

The 1985 Firebird dashboard was more rounded. Shown is the Trans Am edition with optional Lear Siegler front buckets, which featured six areas of adjustment including lumbar support. They came with Pallex cloth trim or, as shown here, in leather with Pallex cloth side panels. This leather-wrapped Formula steering wheel was a no-cost S/E option, and $75 extra in Firebirds and Trans Ams.

The standard floor console (A) provided storage and a place to locate switches for power windows and mirrors. Full instrumentation (B) was at extra cost on base Firebirds only. When equipped with the optional Tuned-Port Injection fuel-injection setup (C) the 1985 Trans Am 5.0-liter V-8 produced 210-horsepower at 4,400 rpm. An overhead console (D) with reading lamp, flashlight, and storage pocket was standard in S/E, optional in other models.

UPC Code	Description	Retail Price ($)
D42	Cargo security screen	69.00
V08	Heavy-duty cooling system (na w/L69)	
	w/o C60	70.00
	w/C60	40.00
DK6	Interior roof console	
	Firebird, Trans Am	50.00
	Firebird S/E	n/c
K34	Cruise control w/resume, accelerate features	175.00
DX1	Hood appliqué decal (avail. w/Trans Am only)	95.00
A90	Remote control deck lid release	
	Firebird, Trans Am	40.00
	Firebird S/E	n/c
C49	Elec. rear window defogger	145.00
VJ9	Emission requirements for California	99.00
K05	Engine block heater	18.00
W51	Black appearance group (base Firebird only; n.a. w/B57)	
	w/o CC1 (incl. black roof drip molding)	152.00
	w/CC1 (black roof drip molding not incl.)	123.00
B57	Custom exterior group (base Firebird only; n.a. w/W51)	
	w/o CC1 (incl. roof drip & belt reveal molding)	112.00
	w/CC1 (roof drip & belt reveal molding not incl.)	51.00
N09	Locking fuel filter door	
	Firebird, Trans Am	11.00
	Firebird S/E	n/c

UPC Code	Description	Retail Price ($)
U21	Rally gauge cluster w/tachometer, trip odometer	
	Firebird	150.00
	Trans Am and Firebird S/E	n/c
K64	78-amp heavy-duty generator (avail. w/LQ9, LB8 and LG4 w/o C60 only)	
	w/o C49	25.00
	w/C49	n/c
K22	94-amp heavy-duty generator (avail. w/LQ9/LG4 and L69 only, w/C60)	
	w/L69	n/c
	w/LQ9/LG4	25.00
A01	Soft-Ray glass, all windows (req. w/C60)	115.00
CC1	Locking hatch roof w/removable glass panels (na w/BX5/B80)	875.00
BS1	Additional acoustical insulation	
	Firebird, Trans Am	40.00
	Firebird S/E	n/c
C95	Dome reading lamp (na w/DK6)	23.00
TR9	Lamp group	34.00
D27	Luggage compartment lockable load floor	75.00
B48	Luggage compartment trim	48.00
B34	Carpeted front floor mats	20.00
B35	Carpeted rear floor mats	15.00
D35	Sport OSRV mirrors (LH remote, RH manual)	
	Firebird	
	w/o B57 or W51	53.00
	w/B57 or W51	n/c
	Firebird S/E, Trans Am	n/c
DG7	Sport electric OSRV mirrors (LH and RH)	

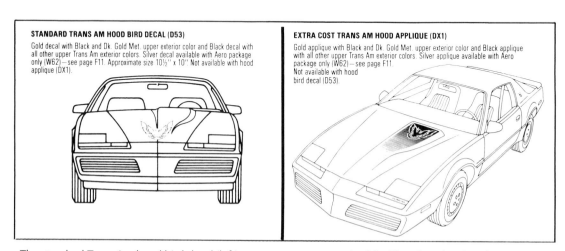

STANDARD TRANS AM HOOD BIRD DECAL (D53)
Gold decal with Black and Dk. Gold Met. upper exterior color and Black decal with all other upper Trans Am exterior colors. Silver decal available with Aero package only (W62)—see page F11. Approximate size 10½'' x 10'' Not available with hood appliqué (DX1).

EXTRA COST TRANS AM HOOD APPLIQUE (DX1)
Gold appliqué with Black and Dk. Gold Met. upper exterior color and Black appliqué with all other upper Trans Am exterior colors. Silver appliqué available with Aero package only (W62)—see page F11. Not available with hood bird decal (D53).

The standard Trans Am hood bird decal (left) came in gold or black. A silver version was available only on cars with the Aero package. Color choice depended upon upper body color, too. The extra-cost venetian blind hood appliqué (right) was again available in the same three colors, with the same restrictions.

UPC Code	Description	Retail Price ($)
	Firebird	
	w/o B57 or W51	139.00
	w/B57 or W51	91.00
	Firebird S/E, Trans Am	91.00
D34	RH visor vanity mirror	7.00
B84	Vinyl bodyside moldings (na w/D84)	
	Firebird, Trans Am (black)	55.00
	Firebird S/E (color-keyed)	n/c
B93	Door edge guards	
	(Firebird w/o W51; bright)	15.00
B91	Door edge guards	
	Firebird w/W51 (black)	15.00
	Firebird S/E, Trans Am (black)	15.00
B80	Roof drip moldings (na w/CC1; Firebird)	
	w/o B57 or W51	
	(bright; incl. belt reveal molding)	61.00
	w/B57 or W51	n/c
BX5	Roof drip moldings	
	(Firebird S/E, Trans Am; black; na w/CC1)	29.00
D84	Two-tone paint (base Firebird only)	205.00
J65	Front & rear power disc brakes (na w/base Firebird)	
	w/o WS6	179.00
	w/WS6	n/c
AU3	Power door locks	130.00
AC3	Six-way power driver seat	225.00
A31	Power windows	
	Firebird, Trans Am	
	w/o B20	195.00
	w/B20 (incl. door map pockets)	225.00
	Firebird S/E (incl. door map pockets)	225.00
UK4	AM/FM ETR stereo radio w/seek-and-scan	168.00
UM7	AM/FM ETR stereo radio w/seek-and-scan, clock	207.00
UM6	AM/FM ETR stereo radio system w/cassette, seek-and-scan, clock	329.00
UX1	AM stereo/FM stereo ETR radio system w/cassette, seek-and-scan, graphic equalizer clock	479.00
UT4	AM stereo/FM stereo ETR radio system cassette, seek-and-scan, graphic equalizer, clock, touch control	519.00
UL5	AM radio, delete (credit)	(56.00)
UQ7	Subwoofer speaker system (avail. w/opt. radios only)	150.00
U75	Power antenna	65.00
AR9	Bucket seats (Firebird and Trans Am)	
	Sierra vinyl trim	n/c
	Genor cloth trim	30.00
WS6	Special Performance pkg. (avail. w/Trans Am only)	664.00

UPC Code	Description	Retail Price ($)
D80	Rear deck spoiler	
	Firebird, Firebird S/E	70.00
	Trans Am, W51	n/c
D81	Aero Wing rear deck spoiler (avail. w/Trans Am only)	199.00
NP5	Leather-wrapped Formula steering wheel	
	Firebird, Trans Am	75.00
	Firebird S/E	n/c
N33	Tilt steering wheel	115.00
D98	Vinyl sport stripes (Firebird only; n.a. w/D84)	75.00
DE1	Louvered rear window sunshield, hinge-mounted (na w/C25)	210.00
Y99	Rally tuned suspension (req. w/QYZ or QYH; Firebird req. w/PE1, N24 or N90)	
	Firebird, Firebird S/E	30.00
	Trans Am	n/c
B20	Luxury trim group (incl. luxury front & rear seats, luxury doors, color-keyed seat belts, split folding rear seat, map pocket, carpeted cowl kick panel)	
	Luxury reclining bucket seats (Pallex cloth trim)	
	Firebird	349.00
	Trans Am	359.00
	Firebird S/E	n/c
	Lear Siegler bucket seats, adjustable (Pallex cloth trim)	
	Firebird	749.00
	Trans Am	759.00
	Firebird S/E	400.00
	Lear Siegler bucket seats, adjustable (leather/Pallex trim)	
	Firebird	1,294.00
	Trans Am	1,304.00
	Firebird S/E	945.00
	Recaro bucket seats, adjustable (Pallex cloth trim)	
	Firebird	985.00
	Trans Am	995.00
	Firebird S/E	636.00
P06	Wheel trim rings (Firebird)	39.00
P02	Five-port wheel covers (Firebird)	39.00
N91	Wire wheel covers w/locking pkg. (na w/Y99; Firebird)	199.00
PE1	14 in. cast aluminum diamond-spoke wheels	
	Firebird	325.00
	Firebird S/E	n/c
N90	15 in. cast aluminum diamond-spoke wheels (req. w/Y99 and QYZ or QYH)	
	Firebird	325.00

UPC Code	Description	Retail Price ($)
	Trans Am	n/c
	Firebird S/E	n/c
N24	15 in. cast aluminum deep-dish hi-tech turbo wheels (req. w/Y99 and QYZ or QYH)	
	Firebird	325.00
	Trans Am	n/c
	Firebird S/E	n/c
N96	16-inch cast aluminum hi-tech turbo wheels (req. w/WS6; Trans Am)	n/c
PB4	Wheel locking package (avail. w/PE1, N24, N90)	16.00
C25	Rear window wiper/washer system (na w/Trans Am)	125.00
CD4	Controlled-cycle windshield wiper system	50.00
QMW	195/75R14 blackwall steel belted radial tires (Firebird)	n/c
QMX	195/75R14 whitewall steel-belted radial tires (Firebird)	62.00
QMY	195/75R14 white-letter steel-belted radial tires (Firebird)	84.00

UPC Code	Description	Retail Price ($)
QHX	205/70R14 blackwall steel-belted radial tires (na w/Y99)	
	Firebird	58.00
	Firebird S/E	n/c
QHW	205/70R14 white-letter steel-belted radial tires (na w/Y99)	
	Firebird	146.00
	Firebird S/E	88.00
QYZ	215/65R15 blackwall steel-belted radial tires (w/LB8, LG4 only)	
	Firebird (W/Y99 only)	132.00
	Firebird S/E (w/Y99 only)	74.00
	Trans Am	n/c
QYH	215/65R15 white-letter steel-belted radial tires (w/B8 and LG4 only)	
	Firebird (w/Y99 only)	224.00
	Firebird S/E (w/Y99 only)	166.00
	Trans Am	92.00
QAC	235/60VR15 blackwall steel-belted radial tires (avail. w/Trans Am only, w/L69, LB9)	177.00
QDZ	245/50VR16 blackwall steel-belted radial tires (avail. w/Trans Am only, w/WS6)	n/c

1986

For 1986, the first-level Firebird featured an all-new taillight design more distinctive than that of the other models. Standard equipment features were expanded to include Rally tuned suspension, 15x7-inch Rally II wheels, a rear deck lid spoiler, body-color sport mirrors, P215/56R15 steel-belted blackwall tires, and a new lower-body accent paint treatment with sport stripe. Monotone finish was also available as a delete option.

Several new exterior paint colors were available, and all 1986 Firebirds featured basecoat/clear-coat paint treatments for an improved, deep-finish appearance. Also new were redesigned seats, wheels with new gold-accent colors, a high-mounted stoplight, optional backlighted gauge cluster with new graphics, and optional electronic day/night inside rearview mirror.

Standard powerplant for the first-level Firebird was the 2.5-liter EFI four-cylinder with five-speed manual transmission. Options included the 2.8-liter multi-port fuel-injected V-6 and the 5.0-liter, four-barrel V-8. A four-speed automatic transmission was available with all but the base powerplant. The standard rear axle was geared at 3.73:1.

The 1986 Firebird S/E featured the back-lighted instrument cluster, redesigned door map pocket, and improved Y99 suspension package with 34-mm front and 23-mm rear stabilizer bars, 12.7:1 power steering ratio, 15x7-inch high-tech cast-aluminum wheels, specifically rated shocks and springs, and P215/60R15 tires.

The base Trans Am engine was the 160 net horsepower 5.0-liter, four-barrel V-8 available with either a five-speed manual or a recalibrated four-speed automatic transmission (also including a revised shift cable mechanism for improved response). The 5.0-liter TPI V-8, rated for 210 net horsepower, again came only with the four-speed automatic gearbox attached.

A lightweight aluminum spare wheel was standard for Trans Ams with the optional engine. In addition, the WS6 suspension package was upgraded to include larger (36-mm) front and (24-mm) rear stabilizer bars, P245/50VR15 Gatorback tires, four-wheel disc brakes, and a limited-slip rear axle. There were 13 base-coat/clear-coat paint finishes available for the top Firebird.

Early Third-Generation Firebird: Strong and Weak Points

Although a variety of minor details were changed from year to year, the early third-generation 1982–86 Firebirds had many basic similarities. Some of these are discussed here, and some opinions about future collectibility are formed.

Comfort

Driver and passenger accommodations in the 1982–86 Firebirds are slightly better than average for similar cars of the same vintages. The seats are low to the floor, but comfortable, especially with optional lumbar-support seats. Positioning of the controls and pedals is very good, although the steering wheel has a slight left-hand offset which takes a little getting used to. Also, in some models, the steering wheel spokes may block the view of the volt meter and oil pressure gauge.

Front compartment passengers have plenty of room, but the rear seat is cramped. Ride quality is extremely firm and solid, but very acceptable for a car of this type. The interior noise level tends to be high, at least through 1984 models. Noise and vibration are particularly noticeable in cars with T-tops. For this reason, I have not allowed any premium for this option except in limited-edition cars that came only with T-tops. The T-top option enhances used car values, but collectors who are purist performance fanatics will likely prefer coupes without the hatch roof treatment.

Much of the high noise level inside Firebirds can be attributed to performance exhaust systems and high-performance tires. It's unlikely that the collector or enthusiast buyer will consider this a drawback, although the average owner could look at things that way. The 1982–86 Firebird heating and ventilation system was very highly rated by testers. One thing noticed by several critics was that the front end sheet metal (and plastic fascias) dropped off dramatically. The drastic slope made

The 1986 first-level Firebird sported a new stripe treatment. Under the hood was an improved 2.5-liter Tech IV (i.e., Iron Duke) engine. The name Firebird again appeared on the front fender behind the wheelhousing.

it hard for first-time drivers to judge the size of the Firebird when parking. (This is something that you will get used to after purchasing one of the cars.) Some owners felt that the pop-up headlight blocked the driver's forward view a little too much.

Convenience

Most experts, even those who are Firebird enthusiasts, have to admit that the convenience level of the 1982–86 models is just average. The low roof and low seats make ingress and egress difficult. The cargo bay is somewhat odd-shaped, so you might need to pick your luggage carefully to fit the available space. However, there is much more total cargo area than in earlier Firebirds, and there are convenient, lockable stowage bins in the floor. Underhood room is minimal and this restricts service access, especially in the later cars with fuel-injection hardware. (Most modern cars share the same problem; they are not designed to encourage do-it-yourself maintenance.)

Workmanship

The level of workmanship in 1982–86 Firebirds was largely improved from that in the earlier models. The new body seemed more solid than in

Artwork in the 1986 sales catalog depicted the front of the Firebird S/E with hatch roof. Shown from the rear was the upgraded base Firebird. A

Firebird S/E decal again decorated the lower-front corner of the door for model identification.

the past, and panel fits were extremely good. Also rated very good were paint finish and trim features, which then got even better from year to year. The 1986 base-coat/clear-coat paint system virtually eliminated all problems in this area, but finish on all the early third-generation models was far above average. This was particularly appreciated by repeat Firebird owners who had experienced paint problems on earlier models.

Interior finish on the 1982 and 1983 cars, especially, was another story. There were problems with plastic dash components and rattling consoles. Some buyers felt that the consoles on these cars had a cheap, flimsy appearance, too. These problems were corrected by the refinements made in the 1984–86 interiors.

Performance

Except in the area of fuel economy, the early third-generation Firebirds drew a very good rating in acceleration, braking, and drivability. In addition, road handling was considered excellent. Some magazines compared the new Firebirds to more exotic and expensive imports, but they were still outstanding high-performance cars for the period in which they were built. And the acceleration and top speeds improved with each passing year.

Future Collectibility

There are several 1982–86 Firebird models that look like good investments for the future.

These are mainly Trans Ams with special interior or powerplant options. I also include the standard 1982 Trans Am in this category because of its first-year-model status.

Two other models have been classified as very good investments for the future. Both are special packages: one (the 1983 Silver Anniversary) having limited-edition characteristics; and the other (the 1984 coupe with Aero package and HO V-8) offering first-year-model status and a special engine. Both are high in demand and limited in supply.

A five-star (excellent) investment rating has been given to one 1982–86 Firebird—the 1984 15th Anniversary Trans Am. While this car had its shortcomings in a technical sense, it also featured a unique appearance, special engine, special-interest status, and low production run. I agree with Cliff Gromer that this car would be a good item to sock away and take to an auction in a few years.

Taking an extremely long-range look, I have also rated six 1985–86 models with three or more stars. This was virtually a necessity, as the lower-level models of these years were made more comparable to earlier Trans Ams, and the 1985–86 Trans Ams were then further upgraded. They had more and better standard equipment than the 1982–84 cars.

Finally, there were several aftermarket conversions of early third-generation Firebirds that future collectors may be interested in. The first of

This profile view of the base Trans Am exhibits the high-tech cast-aluminum wheels, which became regular equipment in 1986. Base engine was the 5.0-liter 160-horsepower four-barrel V-8.

these specials was an updated version of the Bandit Trans Am manufactured by Trans Am Specialties of Cherry Hill, New Jersey. The 1982 version of this model was a bit less expensive and less powerful than the previous edition based on the second-generation body. The new Bandit carried a $16,500 price tag, the 5.0-liter Chevy-built V-8 (water-injected and tweaked for 35 extra horsepower), and a Borg-Warner four-speed manual gearbox. The December 1982 issue of *High-Performance Pontiac* clocked the car at 8.4 seconds for 0–60 miles per hour and 15.2 seconds for the quarter-mile. According to Cliff Gromer, production of 200 units per year was the manufacturer's goal.

Another interesting aftermarket special was the Pontiac Trans Am Motor Sports Edition (MSE) coupe built by Mecham Racing, Inc., of Tacoma, Washington. Finished in a red, white, and blue motif, the MSE featured several different levels of high-performance component packages put together by Huffaker Engineering. Also included were upgraded suspension, interior and sound system features, and special wheels, tires, and emblems. A typical limited-edition MSE Trans Am retailed in the $18,000 bracket.

In addition to these two offerings, various companies across the country advertised a wide range of T-top and convertible conversions, aftermarket Aero packages, and special performance modifications. Many of these will probably wind up in the hands of collectors years from now.

Chapter 17

1987–1992

★★	GTA W61/5.7L TPI
★	GTA W61/5.0L TPI
★★	TA 5.7L TPI
★	TA 5.0L TPI
★★	Formula W63/5.7L TPI
★	Formula W63/5.0L TPI
(+ 1/2 star for GTA w/W63)	

Starting with cars built to 1987 specifications, Firebird buyers were offered more models, more performance, more aerodynamic styling, and more quality. These changes make it possible to view 1987–92 Firebirds as models that share certain distinctions from 1982–86 editions. They can be called "late" third-generation models.

At first, the "more" models looked like less, since Pontiac dropped the S/E designation for 1987 Firebirds, leaving just the base-level coupe and the Trans Am. But the former got a new Formula option and the latter got a Grand Turismo Americano (GTA) option, thereby creating two new products that became official models after 1988. In addition, an Indy Pace Car model option was released for 1989 GTAs and a convertible body style was made available in the Firebird and Trans Am levels starting in mid-1991.

More performance for late third-generation cars was a reaction to the "second-fiddle-to-the-Mustang" image that had developed by 1986. This had a negative impact on sales to enthusiasts. To turn this around, Pontiac dropped the four-cylinder Firebird and brought out hotter V-6 and V-8 engines.

Aero was the buzz word for late third-generation appearance changes. This group of Firebirds shares what is known as an aero-wing spoiler, as well as other aerodynamic enhancements. However, looks alone—at least until 1991—would not be enough to break the pre-1986 and post-1986 Firebirds into separate groups. Styling was just a minor part of the third generation gap.

Quality enhancements played a larger role in setting the latest Firebirds apart from their predecessors. With the introduction of the Trans Am GTA in 1987, Pontiac brought a new standard of fit and finish to the Firebird line. The GTA was a luxury version; one piece of factory literature said it was for "slightly older, more affluent buyers." This model option brought a great deal of refinement to the flagship Firebird and its upgrades tended to trickle down to even base models.

From the viewpoint of the collector/investor, the late third-generation cars probably have more appeal as collectibles than the 1982–86 models, even though they aren't as old. For one thing, they are simply better automobiles. Their added power, improved road manners, and higher quality also enhance overall desirability. On top of all this, they are rarer. Production of Pontiac F-cars never topped 100,000 per year after 1986; in fact, by 1990, output was down to a tad over 40,000 units. And there's no doubt that the 1989 Indy Pace Car is one of the most collectible Firebirds ever built.

1987

The S/E was dropped for 1987. Two Firebirds were listed as models in factory references: the Firebird coupe ($10,773) and the Trans Am ($13,673). But to achieve broader coverage of the sporty car market, Pontiac brought back the Formula as a base model option and added the GTA package for Trans Ams.

Engine changes were big news, as Pontiac rolled up its sleeves to combat the Mustang's king-of-the-hill image. The four-cylinder Firebird vanished. A new 2.8-liter multi-port fuel-injected (MPFI) V-6 was standard in the base model.

This view shows off the 1987 Formula's aero-wing rear spoiler and special lower door graphics.

This kammback Firebird was the prototype for the GTA photographed at Pontiac engineering around 1985. The wagon-like rear styling never made it to an assembly line.

The 1987 Trans Am was promoted as Pontiac's "driving machine." Aero body skirting was standard equipment.

Other versions got V-8s, including one borrowed from the Corvette for standard equipment in GTAs.

Each Firebird retained a distinct personality. At the bottom was a notchback coupe stressing low price, value, and efficiency, although its enlarged V-6 could deliver a 0–60-mile-per-hour performance in 11.5 seconds with the five-speed manual transmission. The four-speed automatic transmission was optional, as was a 5.0-liter V-8.

This particular 5.0-liter engine showed up under three different designations: four-barrel V-8; EFI (electronic fuel injected) V-8; and TBI (throttle-body injection) V-8. To confuse matters even more, this motor isn't listed in 1987 sales catalogs, although it is clearly identified in press releases and the factory certified production specialist (CPS) booklet.

Outside, the Firebird coupe had a more integrated exterior appearance achieved by locating its center high-mounted stoplamp in the body-colored spoiler. Inside were new ripple cloth reclining bucket seats with separately adjustable headrests and custom door trim pads. There were new body-colored side moldings and a softer, retuned suspension.

The Formula package was available in two content levels: W61 for $1,273 and W63 (with lots more power goodies) for $1,842. Formulas had a street machine personality with a sculptured, domed hood, body-colored aero-wing spoiler, and Formula graphics. The V-8 (the one with three different names) was standard. With a five-speed, it did 0–60 miles per hour in nine seconds. Also available were a hotter 5.0-liter TPI V-8 or the 5.7-liter Corvette motor.

Other Formula package ingredients included a WS6 performance suspension, P245/50VR16 tires on 16x8-inch high-tech cast-aluminum wheels, and back-lit rally gauge instrumentation.

Regular 1987 Trans Ams represented Pontiac's "driving excitement" cars. They came with a body aero package, aero rear deck spoiler, bird decals, fog lamps, hood air extractors and louvers, neutral density smooth contour taillamps, deep-dish 15x7-inch gold or silver Hi-Tech Turbo wheels, and added gauges. Introduced at midyear was an optional fully articulating Ultima performance seat, which was standard in GTAs.

Powertrain offerings started with the base

165-horsepower V-8. This could be swapped for the 5.0-liter TPI V-8 featuring a high-output cam good for 206 horsepower and 285 foot-pounds of torque. The top option was the 5.7-liter TPI V-8 with roller valve lifters, hardened steel cam, fast-burn combustion chambers, remote-mounted coil, and dual cooling fans. This slightly modified Corvette engine produced 210 horsepower, 315 foot-pounds of torque, and 6.5-second 0–60-mile-per-hour performance. Unlike other engines, it required the four-speed automatic, since the manual gearbox could not handle the torque.

Two content levels were also offered for the Trans Am GTA package: W61, retailing for $1,701, and W63 for $1,958. This new model option was the luxury performance car, with a monochromatic exterior treatment, power-paddle sport mirrors, cloisonné emblems, and gold cross-lace aluminum wheels. The Corvette V-8 with four-speed automatic was standard, along with WS6 underpinnings and a 3.27:1 limited-slip rear axle.

Six monochromatic base coat/clear coat finishes were offered for GTAs, which also came with custom cloth upholstery; back-lit instrumentation; leather-wrapped steering wheel, shift knob, and brake handle; and an electrically driven 140-mile-per-hour speedometer. *Motor Trend's* yearly new car buyer's guide said the GTA

The production version of the GTA was based on the 1987 Trans Am. Upgraded exterior/interior trim and added luxury made this new option package appealing to upscale buyers.

was "so tough, you'd almost expect to find a hot branding iron in the back seat."

Consumer Guide viewed the 1987 Trans Am as a good choice for performance fans and pointed out that the Firebird's 1986 handling improvements now had upgraded performance to complement them. Tight-fitting rear seats and poor handling on icy winter roads were mentioned as faults of the car, but improvements in overall appearance and fit and finish were praised.

In April 1987, *Road & Track* compared a GTA, IROC Camaro, Toyota Supra, and Nissan 300ZX 2 + 2. The magazine stressed that none of these cars offered the most utility, transportation value, or efficiency for the dollar. "They were judged as factory hotrods," it was noted, and the GTA judged well.

At Willow Springs race course, the GTA balanced high speeds with predictable handling. The 0–60-mile-per-hour blast took 7.1 seconds and the quarter-mile in 15.5 seconds at 89.5 miles per hour. Estimated top speed was 145 miles per hour. Trip fuel economy came in at 14.8 miles per gallon.

High ratings were given to steering quickness and feel, front seat comfort, and the high standard of exterior finish. Weak points included an automatic transmission "sometimes needing a gear between second and third," the harsh ride quality, and the existence of rattles and knocks. However, the Firebird and the Camaro were rated the best, depending on what appearance a buyer preferred.

In September 1987, *Road & Track* did another four car test with a 1987 Formula 5.0 liter, 1969 Trans Am, new Mustang LX, and 1970 Mustang Boss 302. The older cars were quickest. The Mustang LX took 6 seconds to go from 0–60 miles per hour and did the quarter-mile in 14.6 seconds at 97 miles per hour. This compared to 6.3 seconds and 14.9 seconds at 92.5 miles per hour for the Formula. An interesting statistic was "observed gross horsepower" (mathematically calculated from performance data): it was 255 for the Formula, versus the factory rating of 205.

While both of the older cars were faster in accelerating, the aerodynamics and higher rear axle gearing of the 1987s gave higher top speeds, plus better handling, braking, and ride quality. And much better fuel economy.

High-Performance Pontiac magazine tested a GTA at 6.2 seconds from 0–60 miles per hour and a 14.2 seconds/94 miles per hour in the quarter-mile. Drivability and handling were highly rated, but the magazine moaned about the absence of a leather interior option.

Model year production of 1987 Firebirds totaled 88,623 cars against 110,483 for 1986. About half of the F-cars built were Trans Ams or GTAs. Model year dealer sales came to 77,635 units, versus 96,208 the previous season.

General Motors boarded up the Norwood, Ohio factory that built F-cars, in August 1987. Production of Firebirds and Camaros was then limited only to the Van Nuys, California, assembly plant. During the year, the corporation also canceled its GM80 program to develop a plastic-bodied, front-wheel-drive F-car replacement. This raised speculation about the future of the Firebird and whether a fourth-generation version would ever see the light of day.

1987 Firebird Specifications

	Firebird	Trans Am
Wheelbase (in.)	101.0	101.0
Length (in.)	188.0	191.6
Width (in.)	72.0	72.0
Height (in.)	59.7	49.7
Front head room (in.)	37.0	37.0
Rear head room (in.)	35.6	35.6
Front leg room (in.)	43.0	43.0
Rear leg room (in.)	28.8	28.8
Front hip room (in.)	56.3	56.3
Rear hip room (in.)	42.8	42.8
Cargo volume (cu. ft.)	12.4	12.4
Fuel tank capacity (gal.)	15.5 (2.8L MFI, 5.0L TPI)	15.5 (5.0L TPI, 5.7L TPI)
	16.2 (5.0L 4 bbl)	16.2 (5.0L 4 bbl)
Curb weight (lbs.)	3,105 (Base)	3,274 (Trans Am)
Standard engine	2.8L MFI V-6	5.0L 4-bbl V-8
Available engines	5.0L 4-bbl. V-8	5.0L TPI V-8
	5.7L TPI	5.7L TPI V-8
Standard transmission	5-speed manual	5-speed manual
Available transmission	4-speed automatic (req'd w/5.7L)	4-speed automatic (req'd w/5.7L)
Standard axle ratio	3.42 w/2.8L 5sp manual	3.23 w/5.0 LG4 5sp manual
	3.42 w/2.8L 4sp automatic	2.73 w/5.0 LG4 4sp automatic
	3.23 w/5.0L LG4 5sp manual	3.08 w/5.0 TPI 5sp manual
	2.73 w/5.0 LG4 4sp automatic	2.73 w/5.7 TPI 4sp automatic
		2.77 w/5.7 TPI 4sp automatic
Standard tires	P215/65R15 BW	P215/65R15 BW
Available tires	P215/65R15 WL	P215/65R15 WL
		P245/50VR16

1988

★★★★	VHO Formula 350
★★	GTA 5.7L TPI
★	GTA 5.0L TPI
★★	TA 5.7L TPI
★	TA 5.0L TPI
★★	Formula 5.7L TPI
★	Formula 5.0L TPI
★	Firebird 5.7L TPI
(+1/2 star for W63)	

For 1988, Pontiac enhanced the performance strengths that had been given to the 1987 Firebirds. The basic offerings remained the same, but each model or option gained some powertrain, trim, and equipment features that underlined its role in the product mix.

Base-priced at $10,999, the entry-level Firebird coupe retained the 2.8-liter multi-port fuel-injected V-6 as standard equipment (the designation for this motor was shortened to MFI). A new 5.0-liter 170 horsepower throttle-body fuel-injected V-8 was available. Other new items included monotone exterior paint, Pallex cloth interior trim, a redesigned four-spoke steering wheel, and a choice of 15x7-inch deep-dish Hi-Tech Turbo or diamond spoke aluminum wheels. All V-8 engines now employed a serpentine accessory drivebelt system.

Pontiac offered two new exterior colors—silver blue metallic and orange metallic—for 1988 Firebirds, plus a camel color cloth trim. There was also an upgrade in instrumentation to a 120-mile-per-hour speedometer with analog gauge clusters, AM/FM stereo sound system with seek-and-scan tuning, and a clock.

The Firebird again represented an economical, but fun to drive car for young-at-heart buyers. Pontiac described it as "distinctive, yet not too much car to handle." The five-speed transmission was standard and the four-speed automatic was optional. For better handling, buyers could get the Y99 Rally tuned suspension at extra cost.

After its well-received revival in 1987, the Formula (still officially an option, but promoted more as a model) came back for 1988 with a healthy new base engine. This 5.0-liter TBI V-8 had five more horsepower. Even better-suited to the car's street machine personality was an optional 5.0-liter TPI V-8. With an improved air induction system, this motor delivered 190 horsepower with the four-speed automatic and 215 horsepower with the five speed. The 5.7-liter, 225-horse TPI V-8 could be ordered, too,

Pontiac's base engine for Firebirds was the 2.8-liter V-6, labeled MPFI in 1987 and MFI in 1988.

Base Firebirds had a new monotone finish for 1988, plus a choice of two wheel designs including these diamond-spoke aluminum rims.

Standard features of the Formula option package were door call-out graphics and high-tech turbo wheels.

but still only with the automatic.

Formulas got the new colors, trims, and speedometer. Formula graphics on the doors were standard, along with 16x8-inch deep-dish Hi-Tech Turbo wheels. The Formula's role in the line-up was to offer the acceleration and handling characteristics of the TA and GTA at a more attractive base price of $11,999, which represented a substantial savings.

To add the Trans Am's sports car visuals, buyers had to part with a minimum of $13,999. Extra standard equipment on all Trans Ams included the 5.0-liter TBI engine, body skirting, aero front end, hood and front fender air extractors, hood air louvers, fog lamps, Soft-Ray tinted glass, and the Y99 suspension.

The standard Trans Am interior was also trimmed in Pallex cloth, although an optional custom trip group was available. It included deluxe door trim and split folding rear seats, unique front bucket seats and Metrix cloth trim.

Popular options for Trans Ams were the TPI V-8s, automatic transmission (5.0 liter only; it was standard on the 5.7 liter), and the WS6 performance suspension.

To upgrade the 1988 Trans Am GTA option package, Pontiac made the W63 package standard.

1988 Firebird Specifications		
	Firebird	Trans Am
Wheelbase (in.)	101.0	101.0
Length (in.)	188.0	191.6
Width (in.)	72.4	72.4
Height (in.)	50.0	50.0
Front head room (in.)	37.0	37.0
Rear head room (in.)	35.6	35.6
Front leg room (in.)	43.0	43.0
Rear leg room (in.)	29.8	29.8
Front hip room (in.)	56.3	56.3
Rear hip room (in.)	42.8	42.8
Cargo volume (cu. ft.)	12.4	12.4
Fuel tank capacity (gal.)	15.5	15.5
Curb weight (lbs.)	3,097	3,309
Standard engine	2.8L MFI V-6 (LB8)	5.0L TBI V-8 (LO3)
Available engines	5.0L TBI V-8 (LO3)	5.0L TPI V-8 (LB9)
		5.7L TPI V-8 (B2L)
Standard transmission	5-speed manual	5-speed manual
Available transmission	4-speed automatic	4-speed automatic (req'd w/5.7L)
Standard axle ratio	3.42 w/2.8L 5sp manual	3.08 w/5.0 TBI 5sp manual
	3.42 w/2.8L 4sp automatic	2.73 w/5.0 TBI 4sp automatic
	3.08 w/5.0L TBI 5sp manual	3.08 w/5.0 TPI 5sp manual
	2.73 w/5.0 TBI 4sp automatic	2.73 w/5.0 TPI 4sp automatic
		2.77 w/5.7 TPI 4sp automatic
Standard tires	P215/65R15 BW	P215/65R15 BW
Available tires	None	P255/65R15 BW

This meant that GTA buyers got just about every top Pontiac convenience option (including AM/FM stereo cassette with graphic equalizer) for $19,299.

The standard 5.7-liter 225 horsepower TPI V-8 with four-speed automatic could move the GTA from 0–60 miles per hour in less than 6.5 seconds. The 5.0-liter TPI engine was again a delete option and could be mated to either transmission, manual or automatic.

The GTA also came with the WS6 suspension, road-gripping fat tires, gold center cross-lace wheels, four wheel disc brakes, 45/45 split folding rear seats with integral headrests, rear deck power pull-down and fully articulating custom front bucket seats with thigh support, power inflatable lumbar support and side bolsters. A 140-mile-per-hour speedometer with analog tachometer and gauges were included, too. Both Trans Am and GTA buyers could add liquid crystal display (LCD) instrumentation at extra cost.

Pontiac's 1988 "Road Cars" deluxe sales catalog was really unusual, with descriptive copy that sounded like an article from *Automobile* magazine. "These three Birds were certainly at the head of the line when Pontiac handed out hot design, technology, and performance for '88." It described the F-car option list as a "Bird Fancier's buffet of power and pizzazz."

Automobile itself had somewhat mixed comments about F-cars, describing them as a "clunky and weighty American GT that still manages to be a kick to drive." In a similar vein, *Motor Trend's 1988 Automotive Yearbook* said, "you can still play kick-the-Porsche with one of the best-performing American anachronisms on the street." Surely, some automotive critics were happy to see real leather seat trim mentioned in the 1988 catalog for GTAs, but others wondered why all of the Firebird refinements didn't include an anti-lock brake system.

Judging by sales figures, the public was also beginning to wonder about the wisdom of buying a Firebird, Formula, or Trans Am. Model year sales at U.S. dealerships declined to 59,459 units. Production totals also continued to shrink, as only 62,467 Pontiac F-cars passed through the gates of the Van Nuys plant. Wags within GM were wondering, too, about whether a replacement for the Firebird and Camaro was needed.

Apparently, the thinking on this changed during 1988, and it was suddenly announced that the GM80 program to develop a rear-drive replacement had been revived. Steve Schneider, publisher of *High-Performance Pontiac,* discussed this with Mike Losh, PMD general manager, early in 1988. "The high-performance market wants a rear wheel drive car. The Mustang and Ford's success with it proves that," said Losh, who targeted a 1991 or 1992 release.

The April 1988 edition of Schneider's magazine included spy photos of the GM80 by Jim Dunne, of *Popular Mechanics.* They showed a coupe that blended some Fiero lines with the look of an imported GT. Dunne predicted a car smaller and lighter than current editions with a 305-ci (5.0-liter) V-8 as the biggest engine.

Additional insight into Firebird future-think evolved in mid-1988, when the Banshee concept

It was hard to tell the difference between a 1987 and 1988 Trans Am, but the deep-dish wheels were nice looking.

Pontiac upgraded the 1988 GTA by making just about every top-of-the-line accessory standard. Otherwise, it was basically unchanged. Continuing to identify late third-generation Firebirds was the aero-wing spoiler introduced the previous season.

car showed up at the GM Teamwork and Technology For Today and Tomorrow exhibit at New York City's Waldorf Astoria hotel.

This Buck Rogers dream car had a fiberglass body on a tubular frame. Its bright red paint and sleek flowing lines provided a performance image that was accentuated by the 4.0-liter, double-overhead cam (DOHC), aluminum V-8 with integral block-head design resting up front. The 230-horsepower engine was linked to a five speed. Inside, the Banshee emphasized high technology and futuristic accommodations.

A more down-to-earth treat for Firebird freaks was an aftermarket-modified Formula 350 concocted by Greg Carroll, an engineer who liked the P-words Pontiac and Paxton. His Carroll Supercharging Company, of Wyckoff, New Jersey, produced the very high-output (VHO) Formula for sale through selected factory dealers in the Garden State. The car's belt-driven Paxton centrifugal blower gave a five-psi pressure boost, raising horsepower to 400 at 2,500 rpm and torque to 475 foot-pounds at 4,000 rpm. *High-Performance Pontiac* featured the VHO, printing a six second 0–60 mph time and 13 second quarter-mile.

The Carroll supercharging package retailed for $8,500 and Carroll's CSC-103H trunk stereo system was a $2,495 option. The conversion voided the normal factory warranty, but Carroll stood behind the powertrain for one year or 12,000 miles, as long as proper service was done. Few VHO 350 Firebirds were built.

1989

★★★★★	20th Anniversary Convertible
★★★★	20th Anniversary GTA
★★★	GTA 5.7L TPI
★★	TA 5.0L TPI
★★	TA 5.8L TPI
★★	TA 5.0L TPI
★★	Formula 5.7L TPI
★	Firebird 5.0L TPI
(+1/2 star for GTA notchback)	
(+ 1 star for actual Pace Car)	

Car collectors of the future will find that the most desirable 1989 Firebird is one without a V-8. The limited-edition 20th Anniversary Trans Am GTA, powered by a turbo-charged and intercooled sequential fuel-injected (SFI) V-6 and selected as the Official Pace Car of the 73rd Indy 500, is actually one of the most collectible Firebirds ever built.

Before focusing on this special-interest model, the regular 1989 Firebird line-up warrants discussion. Four versions of the Pontiac F-car were again offered: Firebird, Formula, Trans Am, and Trans Am GTA. All were now considered separate models.

Every Firebird had more-reliable, foul-resistant Multec fuel injectors, improved rear disc brakes, and Pass-Key anti-theft protection. The entire line now featured base coat/clear coat paint with a new bright blue metallic color offered. Door glass rubber was improved to cut wind noise and seal better. Three-point rear lap and shoulder belts were standard. Options expanded and all models could have T-tops, the full range

The Banshee, seen here at the 1989 Milwaukee Auto Show, was introduced at a New Aero exhibit in mid-1988.

of radios, and a digital CD player with a Delco II theft deterrent system.

Base Firebirds continued to boast of affordable performance. Prices started at $11,999 for the standard 135-horsepower MFI V-6; the 170-horse TBI V-8 was optional. Both of these came standard with a five-speed manual while the four-speed automatic was extra. Firebird V-6s used a new FE1 suspension with 30-mm front and 18-mm rear anti-roll bars. With V-8s the F41 Trans Am suspension was used and air conditioning was standard. Exterior graphics featured a narrower side stripe.

Street performance at a budget price was again the forte of the Formula. Standard engine was the 5.0-liter TBI V-8, but options included both the 5.0-liter, 225 horsepower TPI, and 5.7-liter, 235 horsepower TPI engines. Air conditioning was standard, along with the WS6 suspension. Body side striping was of the same new design seen on base Firebirds. Starting price was up to $13,949.

At $2,500 more, the basic Trans Am had a few GTA upgrades. Engine choices were the same as the Formula's. A limited-slip differential was standard with extra-cost V-8s. The Corvette motor still came only with the automatic transmission. Also standard on the Trans Am was the F41 suspension with 34-mm front and 23-mm rear anti-roll bars and recalibrated spring and shock rates. Firestone Firehawk GTX tires were mounted on 15x7-inch cast-aluminum wheels.

At the top of the line-up, with a $20,339 price tag, was the 5.7-liter GTA. Finally released was an optional new notchback hatch rear window designed to provide a more individual profile and cooler interior. With 330 foot-pounds of torque, the big-engined 1989 GTA was 0.2 to 0.3 seconds faster in the 0–60-mile-per-hour sprint than before.

Buyers who didn't need quite that much power could get the 225-horse 5.0-liter TPI engine with the automatic. And if they preferred driving a stick, they got 10 extra horsepower thanks to a new dual-converter exhaust system. A 3.45:1 final drive axle ratio came with five-speed gearboxes. Limited-slip differentials were standard with all engines.

New deflected-disc valving for gas shocks and struts gave the GTA's WS6 performance suspension a more comfortable ride. It had 36-mm front and 24-mm rear anti-roll bars, P245/50ZR16 Goodyear tires, and lightweight 16x8-inch cross-lace aluminum wheels.

Cloth articulating seats were standard and leather-covered buckets with increased thigh support plus inflatable lumbar and side bolsters were optional, as was a full complement of power and convenience accessories down to a high-tech radio with redundant controls on the wheel hub.

Of course, the big news for 1989 was the 20th Anniversary Trans Am GTA, which Pontiac developed in conjunction with PAS, Inc. The selection of this model as the year's Indy Pace Car marked the third time a Pontiac attained this honor.

This car was the first Pontiac pacer to lead the race without modification from factory specs and the first Trans Am to come with a V-6. The special engine developed 250 horsepower and 340

The 1989 base Firebird was looking more like a Trans Am than ever before. It still lacked the integrated aero-wing spoiler.

New for 1989 was the notchback rear window option for the GTA, as seen here. This had been under development for several years.

foot-pounds of torque, despite having only 3.8 liters of displacement. It had an 8.0:1 compression ratio.

The powerplant was basically the same turbo engine used in Buick Grand Nationals and was dropped into the Firebird with a change in the front crossmember. The air intake system was unique to this new application. PAS also revised the cylinder head porting and pistons for Pontiac.

Because of the motor's cast-iron constitution, only a few internal modifications were required. They included a specific cross-drilled crank, special heads with improved combustion chamber design, higher-flowing exhaust ports, and newly developed pistons. The Garrett T3 turbocharger provided 16.5 psi of boost. An air-to-air intercooler was fitted to cool the intake charge, aiding power production. Special exhaust manifolds were used, along with an electronic spark control ignition and a three-row radiator.

On a test track, the 20th Anniversary Trans Am sprinted from 0–60 miles per hour in 5.5 seconds and covered the quarter-mile in 13.5 seconds at 100 miles per hour. Top speed was in the 150–153-mile-per-hour bracket—about 10 miles per hour faster than a 5.7-liter V-8 Trans Am.

All anniversary Trans Ams were white with camel interiors. Externally, the GTA nose emblem was changed to a 20th Anniversary insignia with similar cloisonné emblems affixed to the sail panels. The normal GTA script on the front fenders was replaced with "Turbo Trans Am" call-outs. Each vehicle came with a complete set of Official Indianapolis 500 Pace Car decals for the doors and windshield. These could be owner or dealer installed. The price tag on the basic car was $32,000.

Production was preset at 1,500 units. Quite a few of these cars were ordered, with the T-tops used on all Firebirds at Indy. There was an immediate shortage of orders for the cars and dealers were soon discounting the window sticker. Magazines began receiving advertisements offering new 20th Anniversary models in the $20,000 range.

A Pontiac press kit was issued specifically for

1989 Firebird specifications

Engine	5.7-Liter TPI	5.0-Liter TPI
Displacement, cc.	5733	5012
Cu. In.	350	305
Bore x stroke, mm.	101.6x88.4	95.0x88.4
In.	4.00x3.48	3.74x3.48
Compression ratio	9.3:1	9.3:1
Horsepower @ rpm	235 bhp @ 4,400	225 bhp @ 4,400 (manual, dual exhaust)
		190 bhp @ 4,000 (automatic)
Torque @ rpm	330 lb-ft @ 3,200 (Formula)	285 lb-ft @ 3,200 (manual)
	330 lb-ft @ 3,200 (GTA, T/A)	295 lb-ft @ 2,800 (automatic)
Fuel delivery	Tuned port injection	Tuned port injection
Fuel recommendation	91 (premium unleaded)	91 (premium unleaded)
	5.0-Liter TBI	**2.8-Liter MFI**
Displacement, cc.	5,012	2,830
Cu. In.	305	173
Bore x stroke, mm.	95.0x88.4	88.9x76.0
In.	3.74x3.48	3.50x2.99
Compression ratio	9.3:1	8.9:1
Horsepower @ rpm	170 bhp @ 4,000	135 bhp @ 4,900
Torque @ rpm	255 lb-ft @ 2,400	160 lb-ft @ 3,900
Fuel delivery	Throttle body injection	Multi-port fuel injection
Fuel recommendation	87 (regular unleaded)	87 (regular unleaded)

T-tops were optional, but all cars appearing at Indianapolis Motor Speedway for Pace Car duties had them. The Pace Car decals were shipped inside the cars for installation if buyers preferred them.

this car with photos, news releases, and a special 20th Anniversary sales booklet. This sales booklet was also distributed at auto shows and became an instantly collectible piece. The news releases noted that three of the cars, randomly selected from the production run, would be used as actual pace cars at the 500 mile race. The only change from production specifications was the fitting of safety lighting equipment required by Indianapolis Motor Speedway.

The author got to drive one of these three cars when he visited Dave Doern, a well-known Firebird lover, at his Chicago shop. Doern had been given the job of installing the strobe safety lights.

When we took the cars out for a photo session, Doern revealed that they had been used for

a publicity tour in which Pontiac PR men drove the cars from a different starting point (New York, California, and Florida) to Indianapolis for the annual pace car ceremony. He said that one car was damaged in an accident and repaired. In addition, the strobe lights had been difficult to install, because of the tightness of space throughout the cars.

During a short ride, we discovered that the pace cars were quick and fast, but they weren't as smooth-running as V-8 powered Trans Ams. There was nothing unique about them, in comparison to other 20th Anniversary models, except the installation of the extra electronic equipment and strobes.

High-Performance Pontiac drove a 20th Anniversary Trans Am at Riverside International Raceway where it hit 115 miles per hour on the track's short straightaway. However, the California heat—105 degrees Fahrenheit—made the car run hot and hampered performance. In December 1989, the same magazine did a dragstrip shoot out between a Turbo Indy Pace Car and a 1965 GTO with Tri-Power and a four-speed manual

The 1989 Formula's new dual-converter exhaust had a rumble that meant business. A domed performance hood, Formula graphics on the doors and aero-wing spoiler were included.

1989 3.8 Liter Turbo V-6 Specifications

The 3.8-liter turbocharged intercooled V-6 is installed on the 20th Anniversary Trans Am.

Engine	3.8 Liter Turbo
Configuration	90-degree OHV V-6
Valve actuation	Pushrods, hydraulic roller lifters
Displacement, cc.	3,791
Cu in.	231
Bore x stroke, mm.	96.52x86.36
In.	3.80x3.40
Compression ratio	8.0:1
Horsepower @ rpm	250 bhp @ 4,000
Torque @ rpm.	340 lb-ft @ 2,800
Fuel delivery	Sequential fuel injection (SFI)
Fuel recommendation	91 (regular unleaded)
Block material	Cast iron
Cylinder head mat'l	Cast Iron
Pistons	Aluminum alloy
Turbocharger	Garrett T3, air-cooled center housing
Intercooler	Garrett 10-tube, air-to-air
Emissions controls	Oxidizing and reducing 3-way monolithic catalytic converter; oxygen sensor; electronic EGR

gearbox. In repeated runs, the new Trans Am pounded the older muscle car, recording one come-from-behind finish at 13.98 seconds and 98.90 miles per hour. After the intercooler and fuel-injection manifold were packed with dry ice to keep the turbo-charger cool, the car turned in four even faster quarter-mile passes; three with terminal speeds over 100 miles per hour. The quickest was 13.74 seconds at 101.80 miles per hour.

Road & Track also tested a Turbo Trans Am. This car moved from 0–60 miles per hour in 5.3 seconds and took 13.91 seconds to cover the quarter-mile. It required 269 feet to stop from 80 miles per hour. Fuel mileage was estimated at a surprisingly low 14 miles per gallon.

Overall, the Firebird remained a racy car when equipped with either of the more powerful V-8s or the turbocharged V-6. However, the standard V-6 gave mediocre performance and the base V-8 lacked real muscle, despite the fact that it emitted deceptively throaty exhaust sounds. Handling was good on smooth roads, but suffered when the car was driven on gravel, bumpy pavement, or ice and snow. Fuel economy was about 16 miles per gallon city and 25 miles per gallon highway for V-8s and around 19–27 miles per gallon for the base V-6.

Low seat cushions and severely raked seat-backs produced an awkward driving position and hampered visibility. Tall drivers were likely to bang their knees on the dashboard shelf, while climbing in or out of the F-car. Headroom was tight, especially in the rear, where ride comfort was terrible. Despite the folding rear seatback, an uneven luggage compartment floor and shallow cargo bin kept storage space minimal.

The Firebird's styling looked great, especially on the low Trans Am. Quality seemed to improve each year, and 1989 was no exception. However, the basic design of the F-car remained impractical and uncomfortable for everyday transportation. Yet, as a factory hot rod, it was one of the best bargains in the late-model muscle car field. The Firebird—especially the TA and GTA—represented the average man's Porsche or Corvette.

This season, sales of the Firebird reversed their steady downward slide. Total production of F-cars during the 1989 model year rose to 64,406 units, about 2,000 more than the previous models. It's likely that this was due to the offering of the 20th Anniversary edition and the excitement it generated among enthusiasts. Actually, 1,550 of the turbo V-6 models were built, including two convertibles that PSA created by adding Camaro parts. T-tops were used on 1,515 of the Pace Cars and 1,350 had leather interiors.

1989 Firebird Models and Base Prices

Model Number	Model Name	Dealer Cost	Suggested Retail Price
S87	Firebird	$10,715	$11,999
S87 with W66	Formula	$12,456	$13,949
W87	Trans Am	$14,287	$15,999
W87 with Y84	(Trans Am) GTA	$18,163	$20,339

Dealer destination charge (all models):	$439

Basic standard equipment: 2.8-liter Multi-Port Fuel-Injected V-6; power front disc brakes; five-speed manual transmission (automatic transmission on GTA); and P215/65R15 black sidewall tires.

1989 Firebird Options and Accessories

UPC Code	Description	Retail Price ($)
Option Packages		
1SA	Firebird V-6 engine. Option package # 1 includes: (C60) air conditioning; B84 body side moldings; (TR9) lamp group, consists of I/P courtesy lamp, luggage compartment lamp, console lamp and tone generator warnings (seat belt, keys and headlamps).	
	Total package $889; savings $1,200. Net price:	(311.00)
1SB	Firebird V-6 engine. Option package # 2 includes Option package # 1plus (A31) power windows; (AU3) power door locks including door map pockets; (AH3) 4-way manual driver seat adjustment; (K34) cruise control; (A90) remote deck lid release; and (DC4) inside rear view mirror with dual reading lamps.	
	Total package $1,587; savings $1,400. Net price	(187.00)
1SA	Firebird with V-8 engine, Formula and Trans Am. Option package #1 includes (B84) body side moldings; (TR9) Lamp group, consists of I/P courtesy lamp, luggage compartment lamp, console lamp and tone generator warnings (seat belt, keys and headlamps); (A31) power windows; and (AU3) power door locks including door map pockets.	
	Total package $499; savings $1,200. Net price:	(701.00)
1SB	Firebird with V-8 engine, Formula. Option package # 2 includes Option package # 1 plus (AH3) 4-way manual driver seat adjustment; (K34) cruise control; (A90) remote deck lid release; (DC4) inside rear view mirror with dual reading lamps; and (DG7) left- and right-hand power Sport Mirrors.	
	Total package $883; savings $1,400. Net price:	(517.00)
1SB	Trans Am/Firebird with V-8 engine, Formula. Option package # 2. Same as option package #2 for Firebird, except no charge for (AH3) power adjustable driver's seat.	
	Total package $848; savings $1,400. Net price:	(552.00)
R6A	Firebird and Formula Value Option package. Includes (CC1) T-top roof; and (UM6) AM/FM stereo with cassette.	
	Total package $1,052; savings $250. Net price:	802.00
R6A	Trans Am Value Option package includes (CC1) T-top roof; (B20) Custom interior trim; (UM6) AM/FM stereo with cassette; and (D42) cargo area screen.	
	Total package $1,414; savings $325. Net price:	1,089.00
R6A	Trans Am GTA Value Option package includes (CC1) T-top roof; and (AQ9) Leather interior(B20).	
	Total package $1,370; savings $350. Net price:	1,020.00
Engines		
L88	MPFI V-6 in Firebird	n/c
L03	5.0-liter Throttle-Body Fuel-Injected V-8. Not available in GTA; no-cost option in Formulas and Trans Ams.	409.00
LB9	5.0-liter MPFI V-8 requires (G80) limited-slip differential. Not available: Firebird	
	Formula/Trans Am	745.00
	GTA (in place of base V-8)	(300.00)
B2L	5.7-liter MPFI V-8 requires (MXO) automatic transmission; (KC4) engine oil cooler; (J65) 4-wheel disc brakes; (G80) limited-slip differential; and (QLC) tires. Not available: Firebird	
	Formula/Trans Am	1,045.00
	GTA	n/c
Transmissions		
MM5	Five-speed manual. Not available with 5.7-liter V-8. Requires J65 4-wheel disc brakes with LB9 engine. Firebird, Formula and Trans Am	n/c
	GTA with LB9 V-8	(490.00)
MXO	Four-speed automatic. Firebird, Formula and Trans Am. Standard: GTA	515.00

UPC Code	Description	Retail Price ($)
C60	Custom air conditioning. Required on Firebird with L03 V-8. Standard on other models. Firebird	795.00
G80	Limited-slip differential. Available only and required with LB9 or B2L V-8s. Standard: GTA Firebird, Formula, Trans Am	100.00
C49	Electric rear window defogger (included on GTA)	150.00
N10	Dual converter exhaust. Required with LB9 V-8 with MM5 transmission on Firebird Formula and Trans Am GTA; required with B2L V-8 on Formula	155.00
KC4	Engine oil cooler. Required with B2L V-8; not available with other V-8s. Standard: GTA	110.00
NB2	Required California emissions	100.00
CC1	Locking hatch roof. Removable glass panels. Includes sunshades.	920.00
TR9	Lamp group. Consists of instrument panel courtesy lamp; luggage compartment lamp; console lamp; and tone generator warnings. Standard: GTA	34.00
D86	Deluxe two-tone paint. Not available: Trans Am or GTA. Standard: Formula. Firebird:	150.00
WX1	Lower-accent-style two-tone paint. Delete option for Formula only.	(150.00)
J65	4-wheel power disc brakes. Available only and required with B2L V-8 or LB9 V-8 with MM5 manual transmission. Standard: GTA. Formula:	179.00
AU3	Power door locks. Available only with A31 power windows. Standard: GTA	155.00
DG7	Dual power remote Sport Mirrors. Standard: GTA	91.00
A31	Power windows. Available only with AU3 power door locks. Includes door map pockets. Standard: GTA	250.00

Radio Equipment

UPC Code	Description	Retail Price ($)
UM7	AM/FM ETR stereo and clock. Consists of Delco electronic seek-and-scan tuning, integral digital clock, co-axial front speakers, and dual extended-range rear speakers. Not available: GTA	n/c
UM6	AM/FM ETR stereo with cassette and clock. Consists of Delco AM/FM ETR stereo and cassette with auto reverse, electronic seek-and-scan tuning, integral digital clock, and dual front and rear co-axial speakers. Not available: GTA	132.00
UX1	AM/FM ETR stereo with cassette, graphic equalizer, and clock. Consists of Delco AM stereo/FM ETR stereo radio and cassette with auto reverse and search-and-replay, and electronic seek-and-scan tuning. Integral 5-band graphic equalizer. Integral digital clock and dual front and rear co-axial speakers. Not available: GTA	
	Firebird, Formula, Trans Am without value package R6A	282.00
	Firebird, Formula, Trans Am with value package R6A	150.00
UT4	AM/FM ETR stereo with cassette, graphic equalizer, clock, and touch control. Consists of Delco AM stereo/FM ETR stereo radio and cassette with auto reverse and search-and-replay, and electronic seek-and-scan tuning. Integral 5-band graphic equalizer. Integral digital clock, touch control for volume, tone, balance and fade. Steering wheel controls. Dual front and rear co-axial speakers. Not available: Firebird. Standard: GTA	
	Formula, Trans Am without value package R6A	447.00
	Formula, Trans Am with value package R6A	315.00
UA1	AM/FM ETR stereo with compact disc player, clock and graphic equalizer. Consists of Delco AM stereo/FM ETR stereo radio and compact disc player, electronic seek-and-scan tuning. Integral 5-band graphic equalizer. Integral digital clock. Dual front and rear co-axial speakers. Not available in Firebird. Standard: GTA	
	Formula, Trans Am without value package R6A	528.00
	Formula, Trans Am with value package R6A	394.00
	GTA	79.00
U75	Power antenna. Standard: GTA	70.00

UPC Code	Description	Retail Price ($)
D42	Cargo security screen. Standard: GTA hatchback	
	Not available: GTA notchback	59.00
Seats		
AR9	Custom reclining bucket seats with Palex cloth trim. Not available: GTA	n/c
B20	Luxury interior trim. Consists of front-and-rear luxury seats and door panels, split folding rear seat, carpeted cowl kick panel, Matrix cloth trim. Trans Am:	293.00
AQ9	Articulating luxury bucket seats. Standard in GTA with Metrix cloth trim. Optional only in GTA with Ventura leather trim.	450.00
AQ9	Notchback roof. GTA:	700.00
Tires		
QPH	P215/65R15 blackwall touring tires. Firebird only.	n/c
QNS	P215/65R15 blackwall steel-belted radial (Not with B2L V-8). Trans Am only:	n/c
QLC	High-performance P245/50ZR16 blackwall steel-belted radial. Not available with Firebird. Includes WS6 performance suspension and PW7 wheels. Standard: Formula, GTA. Trans Am only:	385.00
Wheels		
N24	15-inch deep-dish charcoal Hi-Tech Turbo aluminum wheels with locking package. Not available for Formula or GTA.	n/c
PEO	16-inch deep-dish Hi-Tech Turbo aluminum wheels with locking package. Formula only.	n/c
PW7	16-inch color-coordinated (black, silver or gold) Diamond-Spoke aluminum wheels with locking package. Trans Am and GTA only.	n/c

1990

★★★★	Trans Am 1LE 5.7L
★★★	Trans Am 1LE 5.0L
★★★	GTA 5.7L TPI
★★	GTA 5.0L TPI
★★★	TA 5.7L TPI
★★	TA 5.0L TPI
★★	Formula 5.7L TPI
★	Formula 5.0L TPI

For model year 1990, Pontiac lowered the base retail price of the Firebird coupe, although the three other models had hiked window stickers. An inflatable driver's side air bag restraint system (eliminating redundant radio buttons) was now included for all Firebirds. There were new instrument panel switches for the rear defogger, rear hatch release, and fog lamps. All models now had body-color sport mirrors. Brilliant red metallic was a new body color replacing flame red.

Under the hoods, Tuned Port Injection V-8s had a speed-density metering feature added. The 5.0-liter TPI V-8 replaced the 5.0-liter TBI V-8 as the standard Trans Am engine and a 3.1-liter V-6 was standard, while the turbo V-6 option disappeared.

Things didn't change much on the 1990 Firebirds, as you can tell from this photo of the Trans Am GTA. Pontiac had planned an abbreviated model-year and was concentrating on several significant revisions due for the early 1991 models that arrived in the first quarter of calendar year 1990.

Despite dropping to $11,320, the Firebird got custom seats like the 1989 Trans Am, with new door armrests, plus a new self-adjusting parking brake on all models. The larger V-6 generated five more horsepower at 140 and 12 percent more torque with 180 foot-pounds at 3,600 rpm. Firestone P215/65R15 Firehawk FX tires on 15x7-inch high-tech aluminum wheels (bright-faced with charcoal metallic ports) were regular equipment. The standard rear spoiler was unique to this model. Air conditioning, a cargo screen, and a rear window defogger were optional.

At $14,610 in basic trim, the Formula added the aero-wing spoiler, air conditioning and tinted glass, plus a domed hood and door graphics. Standard again was the 5.0-liter 170 horsepower TBI V-8. Options included the Trans Am's 5.0-liter TPI V-8 (200 horsepower with the four-speed automatic and 225 horses with the five speed manual and dual catalytic converter exhaust) and the GTA's 5.7-liter 235 horsepower TPI V-8, which again came with the four speed automatic only. The WS6 suspension and P245/50ZR16 tires on deep-dish high-tech wheels were

featured. With the TA and five speed or GTA engines, four-wheel disc brakes and a limited-slip differential were available.

Priced from $16,510, the Trans Am had a complete aero body package and the engines previously noted. The F41 suspension and limited-slip differential were included. Standard tires were P215/65R15 Firehawk GTXs on 15-inch deep-dish Hi-Tech Turbo rims. Buyers could add the Formula's 16-inch tires, but on cross-lace wheels painted either silver metallic or black. Four-wheel disc brakes were available with the 5.7-liter engine.

The interior was like the GTA, including leather appointments (although the leather articulating seats could not be ordered for the Trans Am). Air and tinted glass were standard; a rear window defogger and cargo screen were extra.

The GTA continued to set the pace as one of America's finest performance cars. The 5.7-liter engine was standard; the 5.0-liter TPI V-8 could be ordered to get a credit. GTA buyers who wanted to could also get the WS6 suspension, 16-inch Goodyear tires (good for 154 miles per

1990 Firebird Specifications

Engine	5.7-liter TPI	5.0-liter TPI
Displacement, cc	5,733	5,012
Cu. In.	350	305
Bore x stroke, mm.	101.6x88.4	95.0x88.4
In.	4.00x3.48	3.74x3.48
Compression ratio	9.3	9.3
Horsepower @ rpm	235 @ 4,400	225 @ 4,400 (GTA, Formula manual)
		200 @ 4,000 (formula auto, T/A man.)
Torque @ rpm	340 @ 3,200	300 @ 3,200 (GTA, Formula manual)
		285 @ 2,800 (Formula auto, T/A man.)
Fuel delivery	Tuned port injection	Tuned port injection
Fuel recommendation	91 (premium unleaded)	91 (premium unleaded)

	5.0-Liter TBI	3.1-Liter MFI
Displacement, cc.	5,012	3,128
Cu. In.		305 191
Bore x stroke, mm.	95.0x88.4	88.9x84.0
In.		3.74x3.48 3.50x3.31
Compression ratio	9.3:1	8.9:1
Horsepower @ rpm	170 @ 4,000	140 @ 4,400
Torque @ rpm	255 @ 2,400	180 @ 3,600
Fuel delivery	Electronic fuel injection	Multi-port fuel injection
Fuel recommendation	87 (regular unleaded)	87 (regular unleaded)

hour), and 16x8-inch gold cross-lace wheels.

Inside the top-level Firebird were cloth-covered Ultima seats (leather seats and door trim optional), leather-wrapped controls, air, tinted glass, rear defogger, cargo screen, and Delco ETR cassette sound system. Power equipment included windows, door locks, antenna, and side-view mirrors. Split folding rear seats were no longer available, but their replacement had seatbacks that could be folded for more cargo space.

Pontiac coined a new term to describe its typical Firebird customer: "new collar." This market segment was said to be comprised of buyers in service-oriented technical and professional occupations. It was also known as the mid-sport market niche, made up of enthusiasts looking for an American alternative to luxury sport imports. If "mid-sport" sounds a little middle-of-the-road, it probably was. The 1990 Firebirds were enjoyable, but hardly collectible.

There was at least one exception. Buried deep in a factory-issued car distribution bulletin was a 1LE option that evolved out of the factory racing programs. It was a car intended for real enthusiasts who were interested in entering Firebirds or Camaros in SCCA Showroom Stock racing, International Motor Sports Association (IMSA) Firehawk events, or GM of Canada's Motorsports series.

To get a car equipped with the 1LE option, the Trans Am buyer had to start by specifying a special G92 rear axle and either the more powerful 5.0-liter V-8 with five speed or the 5.7-liter V-8, but no air conditioning with either. To this was added the 1LE high-performance group, G80 limited-slip differential, J65 four-wheel disc brakes, KC4 oil cooler, QLC 16-inch tires, N10 dual converter exhaust system, and C41 air-conditioning delete credit. Pontiac then automatically added the 3.42:1 rear axle ratio, aluminum driveshaft, 18-gallon fuel tank with high-wall reservoir and special wide-strainer pick-up, heavy-duty Corvette front and rear disc brakes, WS6 suspension, and specific shock absorber valving.

The 1LE engines included special features. For instance, the 5.0-liter used cast-iron block and heads, a speed-density air/fuel metering system, special idler pulley, and 9.3:1 compression pistons. This produced a 330 foot-pounds of torque at 3,200 rpm. This engine came linked to a close-ratio five-speed manual gearbox and utilized a special fuel delivery system.

Though meant for showroom stock racing, 1LE cars were fully street legal. They could be driven everyday by anyone willing to put up with the harsh ride characteristics created by the taught suspension. The ride quality was particularly harsh on irregular road surfaces likely to be encountered away from a racetrack.

Performance in the 5.0-liter 1LE was brutal, too. The quarter-mile could be covered in 14.47 seconds with a terminal speed of 96.60 miles per hour. While that didn't make this option the fastest ever offered in a Trans Am, it was close to the top. In addition, the package was so difficult to locate and order (most dealers didn't know it existed) that 1LEs are extremely rare.

In fact, all 1990 Firebirds are relatively rare, partly due to a shortened model year. Build-out for these cars took place in mid-December 1989, at the Van Nuys assembly plant. By December 26, Pontiac had issued a press kit on the 1991s, which showed up at the new car shows in January. By late-February 1990, production versions of the so-called "early 1991s" were rolling out of Van Nuys. Calendar-year sales of 1990 models came to just 38,580 cars. Model-year production was 40,376 units, of which about 75 percent were base Firebirds, 16 percent Formulas, and 5 percent Trans Ams or GTAs.

1990 Firebird Models and Base Prices

Model Number	Model Name	Dealer Cost	Suggested Retail Price
S87	Firebird	$10,109	$11,320
S87 with W66	Formula	$13,047	$14,610
W87	Trans Am	$14,743	$16,510
W87 with Y84	(Trans Am) GTA	$20,825	$23,320
Dealer destination charge (all models):		$439	

Basic standard equipment: 3.1-liter Multi-Port Fuel-Injected V-6; power front disc brakes; five-speed manual transmission (automatic transmission on GTA); and P215/65R15 black sidewall tires.

1990 Firebird Options and Accessories

UPC Code	Description	Retail Price ($)

Option packages

1SB	Firebird V-6 engine. Option package # 1 includes: (C60) air conditioning; B84 body side moldings.	865.00
1SC	Firebird V-6 engine. Option package # 2 includes Option package # 1 plus (A31) power windows; (AU3) power door locks including door map pockets; (AH3) 4-way manual driver seat adjustment; (K34) cruise control; (A90) remote deck lid release; and (DC4) inside rear view mirror with dual reading lamps.	1,603.00
1SB	Firebird with V-6 engine, Formula and Trans Am. Option package #1 includes (B84) body side moldings; ((A31) power windows; and (AU3) power door locks including door map pockets.	495.00
1SC	Firebird V-8, and Formula. Option package # 2 includes Option package #1 plus (AH3) 4-way manual driver seat adjustment; (K34) cruise control; (A90) remote deck lid release; (DC4) inside rear view mirror with dual reading lamps; and (DG7) Left- and right-hand power Sport Mirrors.	889.00
1SC	Trans Am. Option package # 2. Same as option package #2 for Firebird.	854.00

Value Option Packages

R6A	Firebird and Formula Value Option package. Includes (CC1) T-top roof; (Not available with B2L V-8) and (UM6) AM/FM stereo with cassette.	
	Total package $1,070; savings $250. Net price:	820.00
R6A	Trans Am Value Option package. Includes (CC1) T-top roof; (Not available with B2L V-8); (UM6) AM/FM stereo with cassette; and (D42) cargo area screen.	
	Total package $1,139; savings $250. Net price:	869.00
R6A	Trans Am GTA Value Option package. Includes (CC1) T-top roof; (not available with B2L V-8) and (AQ9) leather interior.	
	Total package $1,370; savings $350. Net price:	1,020.00

Engines

LHO	3.1-liter MPFI V-6 in Firebird	n/c
L03	5.0-liter Throttle-Body Fuel-Injected V-8. Not available: Trans Am, GTA. Requires C60 custom air conditioning in Formula.	350.00
LB9	5.0-liter MPFI V-8 Requires (G80) limited-slip differential. Not available: Firebird.	
	Formula	745.00
	Trans Am	n/c
	GTA (in place of base V-8)	(300.00)
B2L	5.7-liter MPFI V-8 Requires (MXO) Automatic transmission; (KC4) engine oil cooler; (J65) 4-wheel disc brakes; (G80) limited-slip differential; and (QLC) tires. Not available: Firebird.	
	Formula	1,045.00
	Trans Am	300.00
	GTA	n/c

Transmissions

MM5	Five-speed manual. Not available with 5.7-liter V-8. Requires J65 4-wheel disc brakes with LB9 engine. Firebird, Formula, Trans Am	n/c
	GTA with LB9 V-8	(515.00)
MXO	Four-speed automatic	
	Firebird, Formula, Trans Am.	515.00
	GTA	Std.
C60	Custom air conditioning. Required on Firebird with L03 V-8. Standard on other models. Firebird:	805.00

UPC Code	Description	Retail Price ($)
G80	Limited-slip differential. Available only and required with LB9 or B2L	
	V-8s. Standard: Trans Am, GTA. Firebird, Formula:	100.00
C49	Electric rear window defogger. Included on GTA.	160.00
N10	Dual converter exhaust. Required with LB9 V-8 with MM5	
	transmission on Formula and Trans Am GTA; required with B2L V-8 on Formula.	155.00
NB2	Required California emissions.	100.00
CC1	Locking hatch roof. Removable glass panels. Includes sunshades. Not available with	
	B2L V-8. Included in price of Value Option package. Otherwise:	920.00
D86	Deluxe two-tone paint. Not available for Trans Am or GTA.	
	Standard on Formula. For Firebird:	150.00
WX1	Lower-accent-style two-tone paint. Delete option for Formula only	(150.00)
J65	4-wheel power disc brakes. Available only and required with B2L V-8.	
	Standard: GTA. Formula, Trans Am:	179.00
AU3	Power door locks. Available only with A31 power windows. Standard: GTA	175.00
DG7	Dual power remote Sport Mirrors. Standard: GTA.	91.00
A31	Power windows. Available only with AU3 power door locks. Includes	
	door map pockets. Standard: GTA.	260.00

Radio Equipment

UPC Code	Description	Retail Price ($)
U75	Power antenna. Standard: GTA	70.00
UM7	AM/FM ETR stereo and clock. Consists of Delco electronic seek-and-scan tuning,	
	integral digital clock, co-axial front speakers, and dual extended-range rear speakers.	
	Not available: GTA	n/c
UM6	AM/FM ETR stereo with cassette and clock. Consists of Delco AM/FM ETR stereo and	
	cassette with auto reverse, electronic seek-and-scan tuning, integral digital clock, and	
	dual front and rear co-axial speakers. Not available: GTA	150.00
UX1	AM/FM ETR stereo with cassette, graphic equalizer, and clock. Consists of Delco AM	
	stereo/FM ETR stereo radio and cassette with auto reverse and search-and-replay, and	
	electronic seek-and-scan tuning. Integral 5-band graphic equalizer. Integral digital clock	
	and dual front and rear co-axial speakers. Standard: GTA	
	Firebird, Formula, Trans Am without value package R6A	300.00
	Firebird, Formula, and Trans Am with value package R6A	150.00
U1A	AM/FM ETR stereo with compact disc player, clock and graphic equalizer. Consists of	
	Delco AM stereo/FM ETR stereo radio and compact disc player, electronic seek-and-scan	
	tuning. Integral 5-band graphic equalizer. Integral digital clock. Dual front and rear co-axial	
	speakers. Not available in Firebird. Standard: GTA	
	Formula and Trans Am without value package R6A	526.00
	Formula and Trans Am with value package R6A	376.00
	GTA	226.00
D42	Cargo security screen. Standard: GTA hatchback. Not available GTA notchback.	69.00

Seats

UPC Code	Description	Retail Price ($)
AR9	Custom reclining bucket seats with Palex cloth trim Not available GTA.	n/c
AQ9	Articulating luxury bucket seats. Standard in GTA with Metrix cloth trim. Optional only in	
	GTA with Ventura leather trim. Included with Value Option package. Without Value Option	
	package:	450.00

Tires

UPC Code	Description	Retail Price ($)
QPH	P215/65R15 blackwall touring tires. Firebird only.	n/c
QNS	P215/65R15 blackwall steel-belted radial (Not with B2L V-8). Trans Am only.	n/c
QLC	High-performance P245/50ZR16 blackwall steel-belted radial. Not available with Firebird.	
	Includes WS6 performance suspension and PW7 wheels. Standard: Formula, GTA.	
	For Trans Am only.	385.00

1991

By the spring of 1990, the early 1991 Firebirds were in the showrooms. Pontiac called them "perpetual emotion machines." They had 14 major changes, led by new styling that was backed up by a hot Street Legal Performance (SLP) option. The big news for late-model enthusiasts was the mid-model-year introduction of a convertible offered in Firebird and Trans Am levels.

All Firebirds had restyled front and rear fascias made of body-color resilient Endura thermoplastic and incorporating low-profile turn signals and integral air dams. The frontal appearance, obviously inspired by the Banshee show car, was smoother and softer. It had cat's-eye openings for the front directional/fog lamps and fade-away feature lines on the front spoiler.

Formulas, Trans Ams, and GTAs sported a redesigned rear deck aero spoiler. The center high-mounted stop lamp was repositioned on the hatch. New rear fascias on Trans Ams and GTAs housed restyled taillamps. The Firebird offered a new Sport Appearance option.

Base-priced at $12,690, the Firebird got the year's new styling with low-profile head-lamps. It shared other improvements with other models: upgraded acoustics, a revised shift indicator on the console, new dark green metallic and bright white finishes, and a standard AM/FM cassette stereo with clock. Driver side air bags and Pass-Key anti-theft protection were also standard.

Though positioned to serve up a classic Firebird look in an affordable package that competed with domestic sport coupes, the base model's optional Sport Appearance package offered the Firebird buyer a chance to add the bold new TA/GTA image. This package added fog lamps in the front fascia and the Trans Am's wedge-shaped lateral body skirting. Engines, transmissions, wheels, and tires were unchanged from 1990. As usual, a wide choice of extras was offered.

Prices for Formulas started at $15,530.

A bulging hood and door decals continued to emphasize the Formula's street machine personality. Both Trans Am and Formula offered special Level II suspension components as standard equipment and a Level III package with fatter stabilizer bars and special springs and bushings was available.

The new convertible came in base-level models or with the Trans Am content, as seen here.

Despite the major restyling, they retained their characteristic hood bulge and door decals. Monochromatic finish replaced two-tone paint on Formulas. Both Firebirds and Formulas could be ordered in an exclusive shade of silver blue metallic. Other colors for all models were the new green and white; bright red; black; brilliant red metallic; medium gray metallic; and (except on GTAs) bright blue metallic.

Formula power teams were the same as in 1990, except that the 5.0-liter TPI engine was rated five horsepower higher with either transmission. A Performance Enhancement Group was required with the two highest output motors. Formula underpinnings were top of the line with the WS6 Sport Suspension, 245/50ZR16 Goodyear Eagle GTs, and deep-dish, hi-tech turbo wheels.

Dressed in its new styling, the Trans Am retained its personality with functional hood louvers and air extractors. The front fascia had integral fog lamps and special ducts that channeled cooling air to the front brakes. Prices started at $17,530.

Engine, transmission, and suspension equipment was basically unchanged from 1990. However, the base TA engine got the same 5 boost horsepower that it received when optionally installed in Formulas. There was also a suspension upgrade to F41 components, plus P215/60R16 tires and 16x8-inch diamond-spoke wheels finished in charcoal.

Air conditioning and a four-way manual driver's seat were standard on Trans Ams. There were countless options. However, the hatch roofs were not available on cars with the 5.7-liter engine.

Listing for $24,530, the GTA boasted the best of everything. All the bells and whistles were standard equipment, except hatch roofs (again a no-no with the standard 5.7-liter V-8), a couple of exotic sound systems, and a handful of interior trim and convenience features.

The GTA carried the new aero package with fog lights and brake cooling ducts in the front fascia, the distinctive new side skirts and functional hood louvers and air extractors. Upholstery, drivetrains, underpinnings, and convenience extras were the same as before,

1991 Firebird Specifications

Engine	5.7-liter TPI	5.0-liter TPI
Displacement, cc	5,733	5,012
Cu. In.	350	305
Bore x stroke, mm.	101.6x88.4	94.9x88.4
In.	4.00x3.48	3.74x3.48
Compression ratio	9.3	9.3
Horsepower @ rpm	240 @ 4,400	230 @ 4,400 (GTA, Formula manual, T/A opt.)
		205 @ 4,200 (GTA, Formula, T/A auto)
Torque @ rpm	340 @ 3,200	300 @ 3,200 (GTA, Formula manual, T/A opt.)
		285 @ 3,200 (Formula auto, T/A auto)
Fuel delivery	Tuned port injection	Tuned port injection
Fuel recommendation	91 (premium unleaded)	91 (premium unleaded)

	5.0-Liter TBI V-8	3.1-Liter MFI V-6
Displacement, cc.	5,012	3,128
Cu. In.		305 191
Bore x stroke, mm.	94.9x88.4	89.0x84.0
In.		3.74x3.48 3.50x3.31
Compression ratio	9.3	8.5:1
Horsepower @ rpm	170 @ 4,000	140 @ 4,400
Torque @ rpm	255 @ 2,400	180 @ 3,600
Fuel delivery	Electronic fuel injection	Multi-port fuel injection
Fuel recommendation	87 (regular unleaded)	87 (regular unleaded)

except for the Street Legal Performance option.

But the Firebird convertible deserves the spotlight. "Convertibles will build on Firebird's fun-to-drive image," said Pontiac General Manager John G. Middlebrook, who got his new job in 1990. "They also give us three entries in a growing market in which the Sunbird convertible had a record-setting year in 1990 and is continuing its strong sales in 1991."

The Firebird series had not had a ragtop since 1969 and the Trans Am, of course, had never been offered as a convertible in any type of volume, with only eight such cars produced during its first season. However, it now seemed wise to add both droptops to build niche sales. While total U.S. car sales had dropped since 1988, the convertible segment had climbed from 100,000 (one percent of industry volume) deliveries to 135,000 (two percent of industry) in 1990.

"Pontiac expects to expand that even more in

The 1991 Trans Am coupe had new Trans Am decal call-outs on its wedge-shaped lateral body skirting.

Hatch roofs were not available with the 5.7-liter V-8, which was standard in the 1991 GTA. This car must have the 5.0-liter V-8, which was optional and gave the buyer a $300 reduction in price.

the 1991 model year," announced Middlebrook on January 3, 1990. "The Firebird and Trans Am soft top models will generate showroom excitement with 2,000 units scheduled during the remainder of the model year."

Firebirds and Trans Ams came as convertibles with base prices of $19,159 and $22,980, respectively. Both ragtops had the Sport Appearance Package as standard equipment. The Trans Am also included hood louvers and air extractors. It was available with leather articulating bucket seats for $780 extra.

The fully lined, manually operated convertible tops were available in either black or beige. They stowed neatly beneath a hideaway tonneau cover, leaving no visible signs of the top. Three sturdy latches secured the top and a dash-mounted switch released the hatch cover.

Standard power came from the 3.1-liter, 140-horsepower MFI V-6 with four-speed automatic. The 5.0-liter, 170-horse V-8 could be added to Firebird ragtops. Trans Am versions got the 205-horsepower, 5.0-liter TPI V-8 with five-speed manual or optional four-speed automatic transmission.

After the convertible was released, Pontiac reissued its small 12-page regular sales catalog, replacing the red GTA coupe with a red Trans Am convertible. Copy on the page was changed to read, "Pontiac's don't get any more exciting than Firebird. The wide-open pleasures of the new Firebird and Trans Am convertibles are proof."

The deluxe sales catalog did not show a ragtop, but had copy that read, "If you've ever seen a Pontiac enthusiast out profiling a classic Firebird convertible and wondered, 'If they'd only build that again,' your wait is over."

In addition to convertibles, another 1991 innovation with appeal to the special-interest crowd was the Street Legal Performance package for Firebirds. Sourced from SLP Engineering, of Toms River, New Jersey, by GM's own Service Parts Organization and offered for all TPI-engined Firebirds, the option represented an emissions-legal, original equipment manufacturer (OEM) quality, integrated group of go-fast goodies.

Designed for easy installation, without major modifications, the SLP kit was available from GM dealers through the *GM Performance Parts Catalog*. The complete package gave 50 more horsepower at approximately 5,500 rpm and pushed the torque peak above 2,800 rpm. It promised a one second better 0–60-mile-per-hour time and improved quarter-mile performance by about 1 second or 6 miles per hour. The SLP parts included cast-aluminum high-flow siamesed intake runners, tri-Y stainless-steel tuned-length exhaust headers, a low-restriction stainless-steel exhaust system, revised engine calibration components (Cal-Pacs and PROMS), and a low-restriction cold air induction unit.

Firebird enthusiasts were also happy to hear of the Firebird's return to SCCA Trans Am racing in 1991, after a seven-year hiatus. A car sponsored by EDS—a wholly owned subsidiary of GM—was constructed by Group 44 of Winchester, Virginia. The chassis-building company, owned by ex-Pontiac race car driver Bob Tullius, turned to Ed Pink Racing Engines for powerplant development. The car made its initial appearance at the Dallas Grand Prix, on June 2, piloted by Scott Lagasse of Corvette Challenge fame.

IntelliChoice, Inc., an automotive research firm, named the 1991 Firebird "best overall value in the under $30,000 sports car" category. The selection was based on five-year projections of the values of cars and vans in 20 different categories. It took into account, cost, insurance, depreciation, financing, service, and fuel requirements.

Motor Trend, in its August 1990 issue, test drove a 1991 Formula. The car did 0–60 miles per hour in 6.5 seconds and covered the quarter-mile in 14.8 seconds at 94.6 miles per hour. It required 152 feet to stop from 60 miles per hour and gave 17 miles per gallon in city driving conditions.

High-Performance Pontiac reported on two Firebirds with different SLP Packages. The first car was a 5.7-liter Trans Am, with the same setup offered through dealers. The author drove the car for a week and gave a good summary of its new-for-'91 technology. He compared the bolt-on and plug-in modifications to the famous Royal Bobcat GTOs of the 1960s.

"With 290 [horsepower] at 5,000 rpm and 350 [foot-pounds] of torque living just under your right foot, the temptation is great to roll down the windows, shut off the radio, and flat-foot away from every intersection," he wrote about the Trans Am. "But the car will also drive identically to a standard 5.7 TA until you lean on it."

It was noted that the cold air induction package provided a 30-percent airflow improvement over 1988–90 stock parts and a 40-percent gain over 1985–87 stock parts. The special header system with 1.75-inch primary tubes and three-inch-wide collector pipes increased torque and horsepower from 1,800 to 5,500 rpm. The three-inch diameter tailpipe was dressed up with styled tips. The 0–60-mile-per-hour trip took 5.7 seconds.

The 5.7 TPI Firefox GTA model was put together by Pontiac and PAS, the firm from which the 20th Anniversary Turbo V-6 was sourced. The Firefox was touted as a prototype for a future production option. It produced 330 horsepower with a ZF six speed and went from 0–60 miles per hour in 5.6 seconds. The quarter-mile took 13.38 seconds at 103.5 miles per hour.

While the convertible and the SLP may have helped a bit, 1991 model output probably benefited most from the extended model year. The total released in the fall of 1991 was 50,247 units, which looked like a significant improvement and reversal of the overall late third-generation trends. However, comparative production had actually been off significantly since the start of the run. In January 1991, *Automotive News* had noted double-digit dips in the percent of market share held by GM F-cars and, throughout the year, the ten-day production figures in industry trade journals comparing 1991 to 1990 output for similar periods reflected lower assemblies of Firebirds. F-car sales (Firebirds and Camaros combined) had declined drastically from their peak of 448,413 units in 1978, to 72,895 units for the first 11 months of 1991.

1991 Firebird Models and Base Prices

Model Number	Model Name	Dealer Cost	Suggested Retail Price
S87	Firebird Coupe	$11,332	$12,690
S87 with W66	Formula Coupe	$13,868	$15,530
S67M	Firebird Convertible	$17,109	$19,159
W87	Trans Am Coupe	$15,654	$24,530
W87 with Y84	GTA Coupe	$21,905	$23,320
W67M	Trans Am Convertible	$20,521	$22,980

Dealer destination charge (all models effective Jan. 28, 1991): $490

Basic standard equipment: 3.1-liter Multi-Port Fuel-Injected V-6; power front disc brakes; five-speed manual transmission; and P215/65R15 black sidewall tires.

1991 Firebird Options and Accessories

UPC Code	Description	Retail Price ($)
Option packages		
1SB	Firebird V-6 engine. Option package # 1 includes: (C60) Custom air conditioning; B84 body side moldings; (DC4) Inside rear view mirror with flood lamp.	
	Firebird coupe.	913.00
	Firebird convertible.	890.00
1SC	Firebird V-6 engine. Option package # 2 includes Option package # 1 plus (A31) power windows; (AU3) power door locks including door map pockets; (A90) remote deck lid release; and (DG7) dual power convex Sport mirrors.	
	Firebird coupe total package $1,554, savings $500, net price:	1,054.00
	Firebird convertible total package $1,471, savings $500, net price:	971.00
1SB	Firebird V-8. Option package # 1. Includes (B84) body side moldings. (A31) Power windows, including door map pockets. (AU3) Power door locks. (DC4) inside rear view mirror with flood lamps.	
	Firebird coupe total package $573, savings $500, net price:	73.00
	Firebird convertible total package $550, savings $500, net price:	50.00
1SC	Firebird V-8. Option package #2. Includes same as 1SB Option package #1 plus (AH3) 4-way manual adjustable driver's seat; (K34) cruise control; (A90) remote-control deck lid release; (DG7) Dual power convex Sport mirrors.	
	Firebird coupe total package $984, savings $500, net price:	484.00
	Firebird convertible total package $901, savings $500, net price:	401.00
1SB	Formula and Trans Am. Option Package #1: Includes (B84) Body side moldings; (A31) power windows, including door map pockets; (AU3) power door locks; and (DC4) Inside rear view mirror with flood lamp.	
	Formula/Trans Am coupe total package $573, savings $350, net price:	223.00
	Trans Am convertible total package $550, savings $350, net price:	200.00
1SC	Formula. Option package #2. Includes same as 1SB Option package #1 plus (AH3) 4-way manual adjustable driver's seat; (K34) cruise control; (A90) remote-control deck lid release; (DG7) dual power convex Sport mirrors.	
	Formula coupe total package $984, savings $500, net price:	484.00

UPC Code	Description	Retail Price ($)
1SC	Trans Am. Option package #2. Includes same as 1SB Option package #1 plus (AH3) 4-way manual adjustable driver's seat; (K34) cruise control; (A90) remote-control deck lid release; (DG7) dual power convex Sport mirrors.	
	Trans Am coupe total package $949, savings $500, net price:	449.00
	Trans Am convertible total package $866, savings $500, net price:	366.00

Value Option Packages

UPC Code	Description	Retail Price ($)
R6A	Firebird and Formula Value option package. Includes (CC1) locking T-top roof; (Not available with B2L V-8) and (UX1) AM/FM stereo with cassette and graphic equalizer.	
	Total package $1,070; savings $250. Net price:	820.00
R6B	Firebird and Formula Value option package. Includes (CC1) locking T-top roof; (not available with B2L V-8) and (U75) power antenna; and (D42) cargo area security screen.	
	Total package $1,074; savings $250. Net price:	824.00
R6A	Trans Am Value Option package. Includes (CC1) T-top roof; (Not available with B2L V-8); (UX1) AM/FM stereo with cassette and graphic equalizer; (D42) cargo area screen.	
	Total package $1,139; savings $250. Net price:	889.00
R6A	Trans Am GTA Value Option package. Includes (CC1) T-top roof; (not available with B2L V-8) and (AQ9) Leather interior.	
	Total package $1,395; savings $350. Net price:	1,045.00

Engines

UPC Code	Description	Retail Price ($)
LHO	3.1-liter Multi-port fuel-injected V-6 in Firebird	n/c
L03	5.0-liter Throttle-Body Fuel-Injected V-8. Not available: Trans Am, GTA Requires C60 custom air conditioning in Formula.	350.00
LB9	5.0-liter Multi-Port Fuel-Injected V-8 requires (G80) limited-slip differential and (R6P) Performance Enhancement Group with MM5 manual transmission on GTA and Formula. (Not available in Firebird)	
	Formula	745.00
	Trans Am	n/c
	GTA (in place of base V-8)	(300.00)
B2L	5.7-liter Multi-Port Fuel-Injected V-8 requires (MXO) automatic transmission; (KC4) engine oil cooler; (J65) 4-wheel disc brakes; (G80) limited-slip differential; and (QLC) tires. (Not available in Firebird.)	
	Formula	1,045.00
	Trans Am	300.00
	GTA	n/c

Transmissions

UPC Code	Description	Retail Price ($)
MM5	Five-speed manual. Not available with Firebird with LHO 3.1-liter V-6 or 5.7-liter V-8.	
	Formula and Trans Am	n/c
	In Firebird with L03 V-8 and GTA	(515.00)
MXO	Four-speed automatic	
	Firebird, GTA	n/c
	Formula, Trans Am	515.00
C60	Custom air conditioning. Required on Firebird with L03 V-8. Standard on other models.	
	Firebird	830.00
G80	Limited-slip differential. Available only and required with LB9 or B2L V-8s. Standard: Trans Am and GTA.	
	Formula	100.00
C49	Electric rear window defogger. Included on GTA.	170.00
K34	Cruise control. Requires 1SB Option package; included in 1SC option package.	
	Firebird, Formula, Trans Am	225.00

UPC Code	Description	Retail Price ($)
A90	Remote deck lid release. Requires 1SB Option on Firebird wit LO3 V-8;	
	Included in 1SC Formula and Trans Am Option package. Standard in GTA.	60.00
NB2	Required California emissions	100.00
CC1	Locking hatch roof. Removable glass panels. Includes sunshades. Not available with B2L V-8. Included in price of Value Option package.	
	Otherwise	920.00
DG7	Dual power remote Sport Mirrors. Standard on GTA.	91.00
R6P	Performance enhancement group. Not available for Firebird. Not available with LB9 V-8 with MXO transmission. Required with B2L V-8 with MXO transmission on Formula and GTA. Includes N10 dual converter exhausts, J41/J42 power front disc/rear drum brakes, KC4 engine oil cooler; and GU6 performance axle (with LB9 V-8), GU5 performance axle with B2L V-8, or performance axle ratio. No charge in GTA with LB9 V-8 and MM5 transmission or B2L V-8. Not available: Firebird.	
	Formula, Trans Am	265.00
	GTA with LB9 V-8 and MXO transmission	(265.00)
AU3	Power door locks (Available only and required with A31 power windows. Standard with GTA. Other models:	210.00
A31	Power windows. Available only and required with AU3 power door locks.	
	Includes door map pockets. Standard with GTA. Other models:	280.00

Radio Equipment

UPC Code	Description	Retail Price ($)
U75	Power antenna. Standard on GTA.	85.00
UM6	AM/FM ETR stereo with cassette and clock. Consists of Delco AM/FM ETR stereo and cassette with auto reverse, electronic seek-and-scan tuning, integral digital clock, and dual front and rear co-axial speakers. Not available in GTA.	n/c
UX1	AM/FM ETR stereo with cassette, graphic equalizer, and clock. Consists of Delco AM stereo/FM ETR stereo radio and cassette with auto reverse and search-and-replay, and electronic seek-and-scan tuning. Integral 5-band graphic equalizer. Integral digital clock and dual front and rear co-axial speakers. Not available in Firebird, Formula, and Trans Am.	150.00
U1A	Delco AM stereo/FM ETR stereo and CD player with electronic seek up/down tuning; next and previous CD controls; integral 5-band graphic equalizer; integral digital clock; and dual front and rear co-axial speakers.	
	Formula and Trans Am without value package R6A	376.00
	Firebird, Formula and Trans Am with value package R6A	226.00
	GTA	226.00

Seats

UPC Code	Description	Retail Price ($)
AH3	Manual 4-way adjustable driver's seat. Not available on Firebird with LHO V-6. Available only with (and included in) 1SC Option package on Firebird with L03 V-8. Available with 1SB option on Formula and included in 1SC option on Formula. No charge on Trans Am or GTA. On Firebird and Formula with required equipment:	35.00
AR9	Custom reclining bucket seats with Palex cloth trim. Not available GTA.	n/c
AQ9	Articulating luxury bucket seats. Standard in GTA with Metrix cloth trim. Optional only in GTA with Ventura leather trim. Included with Value Option package.	
	Without Value Option package:	450.00
W68	Sport Appearance Package. Firebird only. Requires 1SB or 1SC Option package with LHO engine. Includes Trans Am front and rear aero fascias, fog lamps and aero-style Trans Am body side moldings.	450.00

Tires

UPC Code	Description	Retail Price ($)
QPH	P215/65R15 blackwall touring tires. Includes N24 wheels. Firebird only.	n/c
QPE	P215/60R16 blackwall steel-belted radial touring tires. Includes PW7 wheels. Not available with B2L V-8. Trans Am only.	n/c

UPC Code	Description	Retail Price ($)
QLC	High-performance P245/50ZR16 blackwall steel-belted radial. Not available with Firebird. Includes WS6 performance suspension and PW7 wheels on Trans Am and GTA; PEO wheels with Formula. No charge Formula and GTA. Trans Am only.	313.00
WDV	Tire warranty enhancement for New York.	65.00
Wheels		
N24	15-inch deep-dish charcoal Hi-Tech Turbo aluminum wheels with locking package. Not available for Formula or GTA.	n/c
PEO	16-inch deep-dish Hi-Tech Turbo aluminum wheels with locking package. Formula only.	n/c
PW7	16-inch color-coordinated Diamond-Spoke aluminum wheels with locking package. Charcoal on Trans Am. Gold on GTA.	n/c

1992

★★★★★	Formula Firehawk
★★★	GTA 5.7L TPI
★★	GTA 5.0L TPI
★★★★	TA Convertible 5.0L TPI
★★★	TA 5.7L TPI
★★	TA 5.0L TPI
★★	Formula 5.7L TPI
★	Formula 5.0L TPI
★★★	Firebird Convertible V-8
★★	Firebird Convertible V-6

(+1/2 star for SLP kit)

Pontiac's F-body 2+2 sport car returned in 1992 for its 25th year with new interior and exterior colors, structural improvements, better weather protection, revised radio graphics and controls, and non-asbestos brake pads. Wheel options were expanded.

Models were now grouped into two series: Firebird and Trans Am. The Firebirds included the base coupe, convertible, and Formula; Trans Ams were the coupe, convertible, and GTA. To excite enthusiasts, Pontiac and SLP Engineering teamed up to make the hot Formula Firehawk available through the factory dealer network as a 1992 special model.

Body changes were subtle. Dark jade gray metallic, dark aqua metallic, light blue, and Jamaica yellow were new paint colors. Structural adhesives and welding improvements were used to develop tighter-fitting body panels and both convertibles got added reinforcement below the B-pillar and extra structural members along their rockers. Silicone-impregnated weather stripping

For 1992, the Formula was made part of the Firebird series, although it had the same equipment as before. It remained Pontiac's bargain-priced, high-performance F-car.

Both Firebird and Trans Am convertibles got structural improvements for 1992, although they looked identical to 1991s. This is the Trans Am model.

was added to door and rear hatch seals. An instrument panel isolation pad significantly reduced interior noise levels.

The addition of a pinhole to the T-top design (replacing a dam system) made the hatch roof more waterproof. Consequently, for the first time since 1989, T-tops became available with the 5.7-liter engine on Formulas equipped with black cross-lace wheels. Firebirds got a new 15-inch gold crosslace wheel option with P215/65R-15 steel-belted black sidewall touring tires and gold nameplates and decals.

New beige upholstery replaced camel on GTAs and was offered in leather or cloth. A graphite color instrument panel had full instrumentation with a tachometer, trip odometer, and Delco ETR AM/FM stereo with cassette and clock. A four-spoke tilt steering wheel with comfort pads at the two and ten o'clock positions was also standard in all models.

Engines were carried over from 1991, except for the one used in Firehawks. The R6P Performance Enhancement Group (oil cooler, dual converter exhausts, four-wheel disc brakes, and performance axle ratio) was required with the 5.7-liter TPI V-8 and available with the 5.0-liter TPI V-8 with five speed. The Firebird coupe and convertible came with Pontiac's Level I suspension; GTAs and Formulas got Level III underpinnings.

Prices for the Firebird series started at $12,505 for the coupe, $19,375 for the convertible, and $16,205 for the Formula. Trans Ams retailed at $18,105 for the coupe, $23,875 for the ragtop, and $25,880 for the GTA.

With the Firebird now taking the lion's share of

sales, Pontiac pushed the fact that the W68 Sport Appearance Package (standard on convertibles) combined with the N90 gold wheel option made it look like a GTA. The coupe and ragtop were still based with V-6 power and the Formula came with the 5.0-liter TBI V-8, WS6 suspension, 16x8-inch wheels, and Z-rated performance tires.

Trans Ams got a different 52P gold crosslace wheel option and gold decal package, but 16x8 diamond-spoke rims with touring tires were standard. The 5.0-liter TPI V-8 was again base equipment, as was the Rally tuned suspension. Convertibles included the same features as coupes. GTAs got the 5.7-liter TPI V-8, WS6 suspension, and the 16x8-inch wheels, but with 16-inch Gatorback tires. All Trans Am goodies were standard in GTAs, plus a power antenna, resume speed control, cargo screen, ISRV mirror with map lamp, articulating bucket seats, and the Performance Enhancement Group.

Profiles of 1991 Firebird buyers revealed quite a bit about the appeal of each model's personality. They looked as follows:

	Firebird	Formula	Trans AM	GTA
Male	45%	66%	65%	78%
Married	52%	45%	45%	34%
Average Age	32	32	32	38
Average Income	$45K	$50K	$58K	$60K
College Grad.	24%	30%	42%	40%
Professional	38%	42%	56%	52%

Other interesting facts revealed in a 1992 Pontiac press kit included a 68-percent decline in national thefts of Firebirds after the Pass-Key II system was introduced as a GTA-only option in 1988 and later installed in all models. During 1988, there were 1,689 current Firebirds reported stolen, but the number had dropped to 526 in calendar 1989. "Many insurers offer a substantial discount on comprehensive coverage for vehicles with Pass-Key," said the news release. The press kit also confirmed production of 2,000 convertibles in 1991 and documented that 38,580 Firebirds had been delivered in calendar year 1990, accounting for six percent of Pontiac sales and four percent of sport segment sales.

A car certain to generate even fewer sport segment sales for 1992 was the Formula Firehawk. Only 250 of these instant collector cars were to

The fanciest Firebird remained the GTA, although the Trans Am ragtop was more expensive. For 1992, a specific Formula model with the 5.7-liter engine was available with T-tops, but GTA buyers had to opt for a 5.0-liter delete option to get a hatch roof.

be built. Boldly highlighted on the cover of the October 1991 *High-Performance Pontiac* magazine was a red coupe under the banner "Quickest Street Pontiac Ever!"

According to the road test, the Formula Firehawk did 0–60 miles per hour in 4.6 seconds and had a top speed of 160 miles per hour. The standing quarter-mile took 13.20 seconds to travel, with the car going 107 miles per hour by the time it was finished.

Representatives of Pontiac and SLP Engineering exhibited one of these cars at the Boston World of Wheels during the first week of 1992. They explained that Firehawks came in street and competition versions. The street version produces 350 horsepower at 5,500 rpm and 390 foot-pounds of torque at 4,400 rpm.

The factory-built Formulas that the Firehawk is based on carried the 5.0-liter TPI V-8, four-speed automatic (both upgraded to 1LE specs) plus air conditioning. Externally, the main differences from stock are five-spoke aluminum-alloy wheels and a Firehawk decal on the right-hand side of the rear fascia. The base car carries an MSRP of $19,242. Adding the Firehawk street package costs $20,753 for a total of $39,995, while the competition package inflates this cost by another $9,905 for $49,990 total. After the donor car is assembled in Van Nuys, Pontiac shipped it to SLP Engineering, in Toms River, New Jersey.

The aftermarket partner company extracted the complete drivetrain and went to work adding a Corvette ZF six-speed gearbox with computer-aided gear selection (CAGS), a Dana 44 rear axle with 3.54:1 gears, a shortened input shaft, and a lightweight 16-pound flywheel. A heavy-duty block fitted with four-bolt mains got a forged steel crank, Gen II cast pistons, Ed Pink Racing connecting rods, and a hydraulic roller cam. Bolted to this are aluminum heads with two-inch intake and 1.56-inch exhaust stainless-steel valves.

A special downdraft port injection manifold is also employed. Designed by Ray Falconer, it features 11 1/2-inch runners, 52-mm throttle-body, high-flow dual filter system, and stainless-steel headers and exhausts with dual catalytic converters.

Suspension modifications include revised spring rates, lowered ride height, new struts, new rear shocks, larger front and rear antiroll bars,

special bushings, and Corvette 11.85-inch disc brakes. Firestone 275/40ZR17 Firehawk tires were mounted on the 17x9.5-inch Ronal wheels. Recaro seats were extra for $995 and the center console inside the car was modified to give more space for shift throws.

The competition versions of the Firehawk package include the Recaro seats, 13-inch Brembo vented disc brakes with four-piston calipers, a roll cage and an aluminum hood. The rear seat was deleted.

SLP Engineering targeted production of five units per week, starting in July 1991, but only on a build-to-order basis. A preproduction example of the competition version was entered in the Bridgestone Petenza Supercar Series race at Lime Rock on May 27, 1991. It took third place and the company soon had plans to make three or four additional racing versions.

The SLP kit was also available through GM Service Parts Organization's 1992 high-performance catalog, but was not shown in sales literature because there was initial uncertainty about 1992 EPA certification. Ultimately, SLP wound up doing only 25 of the $39,995 Firehawks. At least one of these was a Trans Am convertible. The ragtop has SLP serial number 27, but cars 18 and 25 were never built, which explains why there are existing serial numbers higher than the production total.

Despite the obvious Firebird enthusiasm reflected in PMD news releases, model year

Created through a partnership between Pontiac and SLP Engineering, the 1992 Formula Firehawk was the fastest street Pontiac ever produced.

production got off to a sluggish start with only 10,449 cars built through November 1991. *Automotive News* quoted John G. Middlebrook as saying that Firebird deliveries were off six percent.

A month later came GM's historic announcement of massive layoffs and multiple plant closings. *Automotive News* said, "cutbacks at GM could doom the scheduled replacement for the Pontiac Firebird and Chevrolet Camaro and, although Pontiac, Chevrolet, and their dealers would squawk, there is growing sentiment among analysts that the next-generation F-cars are vulnerable."

But once again, though the threat of ending the history of the GM F-cars arose, it just as quickly evaporated. Late in calendar-year 1993, an all-new, fourth-generation Firebird would be brought to market to replace the 25th anniversary 1992 models. Developmental delays would slow the actual introduction of this brand-new car and hold down its 1993 sales. However, by 1994, the yearly output of Firebirds rose to double the 1992 figure and the future of the marque would come to seem secure again.

1992 Firebird Models and Base Prices

Model Number	Model Name	Dealer Cost	Suggested Retail Price
Firebird			
S87	Firebird Coupe	$11,229	$12,505
S67	Firebird Convertible	$17,399	$19,375
S87 with W66	Formula Coupe	$14,552	$16,205
Trans Am			
W87	Trans Am Coupe	$16,258	$18,105
W67	Trans Am Convertible	$21,440	$23,875
W87 with Y84	GTA Coupe	$23,240	$25,880

Dealer destination charge (all models effective Aug. 30, 1991): $490

Basic standard equipment: 3.1-liter Multi-Port Fuel-Injected V-6; power front disc brakes; five-speed manual transmission; and P215/65R15 black sidewall tires.

1992 Firebird Options and Accessories

UPC Code	Description	Retail Price ($)
Option packages		
1SB	Firebird V-6 engine. Option package # 1 includes: (C60) Custom air conditioning; B84 body side moldings; (DC4) Inside rear view mirror with flood lamp.	
	Total coupe price $913, savings $500. Net price:	413.00
	Total convertible price $890, savings $500. Net price:	390.00
1SC	Firebird V-6 engine. Option package # 2 includes Option package # 1 plus (A31) power windows; (AU3) power door locks including door map pockets; (A90) remote deck lid release; and (DG7) dual power convex Sport mirrors.	
	Firebird coupe total package $1,554, savings $750, net price:	804.00
	Total convertible price $1471, savings $750. Net price:	721.00
1SB	Firebird V-8. Option package # 1. Includes (B84) body side moldings. (A31) Power windows, including door map pockets. (AU3) Power door locks. (DC4) Inside rear view mirror with flood lamps.	
	Firebird coupe total package $573, savings $325, net price:	248.00
	Total convertible price $550, savings $325. Net price:	225.00

UPC Code	Description	Retail Price ($)
1SC	Firebird V-8. Option package #2. Includes same as 1SC Option package #1plus (AH3) 4-way manual adjustable driver's seat; (K34) cruise control; (A90) remote-control deck lid release; (DG7) dual power convex Sport mirrors.	
	Firebird coupe total package $984, savings $500, net price:	484.00
	Total convertible price $901, savings $500. Net price:	401.00
1SB	Formula and Trans Am. Option package #1: includes (B84) body side moldings; (A31) power windows, including door map pockets; (AU3) power door locks; and (DC4) Inside rear view mirror with flood lamp.	
	Formula/Trans Am coupe total package $573, savings $350, net price	248.00
	Total Trans Am convertible price $550, savings $325. Net price:	225.00
1SC	Formula. Option package #2. Includes same as 1SB Option package #1 plus (AH3) 4-way manual adjustable driver's seat; (K34) cruise control; (A90) remote-control deck lid release; (DG7) dual power convex Sport mirrors.	
	Formula coupe total package $984, savings $500, net price:	484.00
1SC	Trans Am. Option package #2. Includes same as 1SB Option package #1 plus (AH3) 4-way manual adjustable driver's seat; (K34) cruise control; (A90) remote-control deck lid release; (DG7) dual power convex Sport mirrors.	
	Trans Am coupe total package $949, savings $500, net price:	449.00
	Total convertible price $866, savings $500. Net price:	366.00

Value Option Packages

R6A	Firebird and Formula Value Option package. Includes (CC1) locking T-top roof; (Not available with B2L V-8) and (UX1) AM/FM stereo with cassette and graphic equalizer.	
	Total package $1,064; savings $250. Net price:	814.00
R6A	Trans Am Convertible Value Option package. Includes (AQ9) articulating luxury bucket seats with Ventura leather trim. (UX1) AM/FM stereo with cassette and graphic equalizer.	
	Total package $930; savings $250. Net price:	680.00

Engines

LHO	3.1-liter Multi-port fuel-injected V-6 in Firebird.	n/c
L03	5.0-liter Throttle-Body Fuel-Injected V-8. Not available: Trans Am, GTA. Requires C60 custom air conditioning in Formula.	369.00
LB9	5.0-liter Multi-Port Fuel-Injected V-8 requires (G80) limited-slip differential and (R6P) Performance Enhancement Group with MM5 manual transmission on GTA and Formula. Not available: Firebird.	
	Formula	745.00
	Trans Am	n/c
	GTA (in place of base V-8)	(300.00)
B2L	5.7-liter Multi-Port Fuel-Injected V-8 requires (MXO) automatic transmission; (KC4) engine oil cooler; (J65) 4-wheel disc brakes; (G80) limited-slip differential; and (QLC) tires. Not available: Firebird.	
	Formula	1,045.00
	Trans Am	300.00
	GTA	n/c

Transmissions

MM5	Five-speed manual. Not available: Firebird with LHO 3.1-liter V-6 or 5.7-liter V-8.	
	Formula and Trans Am	n/c
	In Firebird with L03 V-8 and GTA	(530.00)
MXO	Four-speed automatic	
	Firebird, GTA	n/c
	Formula, Trans Am	530.00

UPC Code	Description	Retail Price ($)
C60	Custom air conditioning. Required on Firebird with L03 V-8. Standard on other models.	
	Firebird	830.00
G80	Limited-slip differential. Available only and required with LB9 or B2L V-8s. Standard: Trans Am, GTA	
	Formula	100.00
K34	Cruise control. Requires 1SB Option package; included in 1SC Option package.	
	Firebird, Formula, Trans Am	225.00
A90	Remote deck lid release. Requires 1SB option on Firebird with LO3 V-8;	
	Included in 1SC Formula and Trans Am option package. Standard in GTA.	60.00
C49	Electric rear window defogger. Included in GTA.	170.00
NB2	Required California emissions	100.00
CC1	Locking hatch roof. Removable glass panels. Includes sunshades. Not available with B2L V-8. Included in price of Value Option package.	
	Otherwise	914.00
DG7	Dual power remote Sport mirrors. Standard on GTA.	91.00
R6P	Performance enhancement group. Not available for Firebird. Not available with LB9 V-8 with MXO transmission. Required with B2L V-8 with MXO transmission on Formula and GTA. Includes N10 dual converter exhausts, J41/J42 power front disc/rear drum brakes, KC4 engine oil cooler; and GU6 performance axle (with LB9 V-8), GU5 performance axle with B2L V-8, or performance axle ratio. No charge in GTA with LB9 V-8 and MM5 transmission or B2L V-8. Not available: Firebird.	
	Formula, Trans Am	444.00
	GTA with B2L V-8	n/c
	GTA with LB9 V-8 and MXO transmission	(444.00)
AU3	Power door locks. Available only and required with A31 power windows.	
	Standard with GTA. Other models:	210.00
A31	Power windows. Available only and required with AU3 power door locks.	
	Includes door map pockets. Standard with GTA. Other models:	280.00

Radio Equipment

UPC Code	Description	Retail Price ($)
U75	Power antenna. Standard on GTA.	85.00
UM6	AM/FM ETR stereo with cassette and clock. Consists of Delco AM/FM ETR stereo and cassette with auto reverse, electronic seek-and-scan tuning, integral digital clock, and dual front and rear co-axial speakers. Not available in GTA.	n/c
UX1	AM/FM ETR stereo with cassette, graphic equalizer, and clock. Consists of Delco AM stereo/FM ETR stereo radio and cassette with auto reverse and search-and-replay, and electronic seek-and-scan tuning. Integral 5-band graphic equalizer. Integral digital clock and dual front and rear co-axial speakers. Not available in Firebird, Formula, Trans Am.	150.00
U1A	Delco AM stereo/FM ETR stereo and CD player with electronic seek up/down tuning; next and previous CD controls; integral 5-band graphic equalizer; integral digital clock; and dual front and rear co-axial speakers.	
	Formula, Trans Am without value package R6A	376.00
	Firebird, Formula, Trans Am with value package R6A	226.00
	GTA	226.00
D42	Cargo area security screen. Not available for convertibles.	
	Included with GTA	69.00

Seats

UPC Code	Description	Retail Price ($)
AH3	Manual 4-way adjustable driver's seat Not available on Firebird with LHO V-6. Available only with (and included in) 1SC Option package on Firebird with L03 V-8. Available with 1SB on Formula and included in 1SC on Formula. No charge on Trans Am or GTA. On Firebird and Formula with required equipment:	35.00

UPC Code	Description	Retail Price ($)
AR9	Custom reclining bucket seats with Palex cloth trim Not available GTA.	n/c
AQ9	Articulating luxury bucket seats. Standard in GTA with Metrix cloth trim. Optional only in GTA with Ventura leather trim. Included with Value Option package.	
	Without Value Option package:	450.00
W68	Sport Appearance Package. Firebird only. Requires 1SB or 1SC Option package with LHO engine. Includes Trans Am front and rear aero fascias, fog lamps and aero-style Trans Am body side moldings.	450.00

Tires

QPH	P215/65R15 blackwall touring tires. Includes N24 wheels. Firebird only.	n/c
QPE	P215/60R16 blackwall steel-belted radial touring tires. Includes PW7 wheels. Not available with B2L V-8. Trans Am only.	n/c
QLC	High-performance P245/50ZR16 blackwall steel-belted radial. Not available with Firebird. Includes WS6 performance suspension and PW7 wheels on Trans Am and GTA; PEO wheels with Formula. No charge Formula and GTA. Trans Am only.	313.00
WDV	Tire warranty enhancement for New York	25.00

Wheels

N24	15-inch deep-dish charcoal Hi-Tech Turbo aluminum wheels with locking package. Not available for Formula or GTA.	n/c
PEO	16-inch deep-dish Hi-Tech Turbo aluminum wheels with locking package. Formula only.	n/c
PW7	16x8-inch black Diamond-Spoke aluminum wheels with locking package.	
	All except GTA	n/c
52F	16x8-inch gold Diamond-Spoke aluminum wheels with locking package. Includes gold decals on Trans Am and Trans Am convertible; gold cloisonné emblems on GTA. Available on Trans Am including with beige top on Trans Am convertible. Standard on GTA. Not available on Firebird, Firebird convertible, Formula, Trans Am; Trans Am convertible; and GTA.	n/c

Chapter 18

1993–1998

The introduction of the fourth-generation Firebird, in the fall of 1992, may have saved the nameplate from extinction. The remarkable thing about these Pontiacs (or their counterpart Chevy Camaros) is that they'll catch your eye every time you see one on the streets. Even today, six years after they first appeared, the beautiful lines of these cars are strikingly attractive.

Pontiac was smart enough to back the good looks with a solid safety package, some super performance options, and handling hardware that glued the cars to the road.

They are good cars, too. They're reliable, effi- cient, and about as environmentally "friendly" as a true sports car can be. That's right—a true sports car! With the fourth generation, Pontiac definitely crossed the bridge between "sporty" and "sports." Officially, the Firebirds are four seaters, but you won't want to ride in the back— believe me. When the cars first came out, I wanted one to replace my '84 F-car, but family priorities ruled against the purchase, and I wound up with a sports sedan instead. I think that decision will always hurt a little, and proba- bly more and more with the passage of time.

From a collector's standpoint, the fourth-gen- eration Firebirds are a pleasant balance of factors.

"Fly without ever leaving the ground" is the slogan Pontiac used to promote the fast-standing-still looks of the all-new 1993 "fourth-generation" Fire- bird. The base Firebird coupe can most easily be spotted by its black-finished mirrors and "Firebird" door lettering.

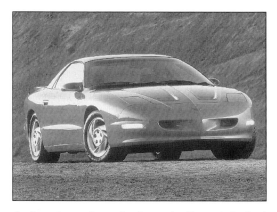

At first glance, from the front, the Formula looked very much like the base Firebird. Its mirrors are body colored, and the doors have no lettering. Instead, there's a "Formula V8" inscription on the right-hand side of the rear bumper. Most other dif- ferences are under the skin.

Peak production seems to be in the 50,000 bracket, which is half what it was in the days when F-cars crested six-figure production totals. However, it's high enough to ensure a good supply of parts out there. That will be important when people start restoring the cars.

Even with five-digit production totals, some option-created models are extremely rare. And if you get into the aftermarket conversions—like the well-known SLP Firehawk—you can find some 1983–88 Firebirds whose population you can count on the tips of your fingers. That's sure to make such cars worth a premium as time goes by.

With this said, let's take a closer look at the fourth-generation Pontiac F-cars.

The fourth-generation Firebirds came out later than other 1993 Pontiacs. Production began in November 1992, at a GM assembly plan in Ste. Therese, Quebec, Canada. The first cars were scheduled to appear in dealer showrooms in early 1993. They actually arrived even later.

Newly styled and engineered, the Firebirds were promoted as Pontiac's "sports cars." They combined eye-catching looks, pirated from Pontiac's 1988 Banshee concept car, with high-performance driving and good roadability.

"Firebird has always symbolized the epitome of Pontiac sports-car-driving excitement," said Pontiac General Manager John G. Middlebrook. "Just like the two million units before it, the fourth generation will carry on that tradition—cars designed to put fun and excitement in the driving experience."

Three models were available: Firebird, Formula, and Trans Am. Ninety percent of the content was new, including a 3.4-liter V-6 with SFI, a 5.7-liter V-8 with MFI, and new dent-, ding-, and rust-resistant body panels. The doors, roof, hatch, and spoiler were made of polymer composite material, and the fenders and fascias were

A '93 Formula Firebird is seen at speed here, and the car could go fast. Pontiac extracted 275 horsepower from the 5.7-liter (350-ci) V-8. It produced 325 foot-pounds of torque at 2,000 rpm. This engine was standard in the Trans Am, too.

The '93 Trans Am came standard with power Sport mirrors, body-color body side moldings, "neutral density" taillamps (also used on Formula), and high-performance P245/50ZR16 steel-belted radial tires.

Reaction Injection Molded (RIM) plastic. Two-sided galvanized steel was used for the rear quarter panels and hood.

The '93 Firebirds had what Pontiac called "a fortified structure." There were several reasons for this: enhanced safety and security, easier servicing, and reduced replacement-parts costs. A bigger motivation was earning reduced insurance rates. "We've improved Firebird's safety features and made it more theft resistant and less costly to repair," said Middlebrook. "And that should help take the bite out of the insurance premiums owners have traditionally paid."

Jack Folden, design chief of Pontiac's Exterior Studio II, said "The car has an international design flavor to it. And yet, it very much says 'I'm an American and proud of it.'" He revealed that new plastic and composite body materials allowed his designers to create more flexible shapes. "It gave us more freedom to twist and flow surfaces together," he said. "We couldn't have gotten that pinched-in, wasp-waisted look without the use of these materials."

Other exterior features included a rakish 68-degree windshield angle, a spoiler with a high-mounted stop lamp built into its center, and a rear hatch with hinge-and-latch hardware. Solar Ray glass was used everywhere, except on T-top roof panels. There were new moldings and emblems, wheels and tires, and rear view mirrors.

"The most aggressive profile this side of Madonna," is how SLP Engineering described its sizzling Firehawk version of the 1993 Trans Am. "This bird has predatory instincts," said Motor Trend of the 3,476-pound, 300-horsepower Firehawk. "Oohs and ahhs are assured."

Kim Willet (left) uses sign language to translate information about the rare 25th Trans Am convertible that speaking narrator Holly McCann (on left) presented at the Chicago Auto Show. This show was used for the official re-introduction of a Firebird ragtop.

Shown here, clockwise from the front, are the 1994 Trans Am Coupe, 1994 Firebird Coupe, 1994 Formula Coupe, and 1994 Trans Am GT Coupe. The Trans Am (in front) had a new front fascia design this year.

Also new were four colors: dark green metallic, yellow, medium red metallic, and gray purple metallic. The performance image of V-8s was enhanced by polished, stainless-steel tailpipe tips projecting from the cars.

The base Firebird coupe was said to be targeted to unmarried, college-educated buyers (mostly female) making over $35,000. It had sporty, sophisticated styling, and a good assortment of performance and safety features.

Also a coupe only, the masculine-looking '93 Formula added high-performance features to the mix. It was targeted at single men, many with college degrees, making over $40,000 per year.

Strong suits of the Trans Am were "unprecedented performance, control, and excitement with a bold, muscular appearance and features required in high-performance sports cars." It was targeted primarily to males who were single or young couples with incomes of $45,000-plus."

Standard Firebird equipment included a front passenger air bag; ABS anti-lock brakes; de Carbon mono-tube shocks; short-and-long-arm front suspension; power rack-and-pinion steering; R-134A air conditioning refrigerant; a locking glove box, and a center-console cup holder (of course). The exterior was characterized by the raked windshield, concealed electric-operated rectangular quartz headlamps, an integrated rear deck lid spoiler, and two one-piece taillamps with multi-colored lenses and amber turn signals. The sleek looks could be upgraded to include body-color body side moldings, a removable hatch roof with an improved locking system, and power outside rear-view mirrors.

The base model also came with newly-styled 16-inch cast-aluminum Hi-Tech wheels, and P215/60R16 touring tires. Power came from a bigger 3.4-liter, 160-horsepower V-6 that delivered up to 28 miles per gallon (highway) with a five-speed manual transmission. Inside were two-way adjustable reclining bucket seats, a Delco ETR AM/FM stereo radio/cassette with four-speaker sound system, cut-pile carpets, rear-folding seat, side window defoggers, and a Pass-Key II theft-deterrent system.

The hot Formula had the same general appearance as the Firebird, except for its body-color Sport mirrors, smoothly-contoured taillamps (with neutral-density lenses), and 16-inch silver sport-cast aluminum wheels with P235/55R16 touring tires. For go power, Pontiac added a 275-horsepower 5.7-liter V-8, six-speed manual transmission, FE2 performance suspension; limited-slip differential; and power four-wheel disc brakes.

The 1993 Trans Am had the same taillamp treatment and wheels as the Formula, but added a specific rear deck-lid spoiler, plus front-and-rear body-fascia panels that included a one-piece resilient front panel and front bumper. Trans Ams also featured fog lamps, and rocker panel extensions. Fat P245/50ZR16 Goodyear Eagle GS-C performance tires were mounted.

Power came from the 275-horsepower at

The 25th Trans Am was announced in January 1994 and went into production the same month. It combined the new-for-'94 nose design with a special blue-and-white appearance package. Only a limited number of these cars were built. Note the distinctive spoiler.

5,000-rpm 5.7-liter V-8, which generated 325 foot-pounds of torque at 2,400 rpm. It came attached to either the six-speed stick shifter, or a four-speed automatic transmission. A performance-calibrated suspension with 3.23:1 axle was standard.

Inside, a 155-mile-per-hour speedometer, rally gauges, and leather steering wheel highlighted the car's aggressive character. A leather shift knob and parking brake handle were also seen. Power windows, power door locks, air conditioning, fully-articulating sport bucket seats,

and a remote-keyless entry system were standard. The standard cloth seats had a six-way-adjustable headrest and adjustable lumbar support pads. A "high-horsepower" 10-speaker sound system with steering-wheel controls was used.

Because of the shortened 1993 model year, only 14,112 Firebirds—of all models—left the assembly line that season. That makes even the factory-issued cars of this vintage rare. In addition, they may be slightly desirable, to collectors, as the first of a new generation.

For collectors who want to go beyond showroom editions, SLP Engineering, of Toms River, New Jersey, again produced an amazingly-potent "Firehawk" version of the Trans Am in 1993. In addition to having twin functional air scoops and special "Firehawk" door decals, it featured a 5.7-liter iron-block V-8 with aluminum heads and a 10.25:1 compression ratio. Generating 300 horsepower at 5,000 rpm and 330 foot-pounds or torque at 2,000 rpm, it came with a choice of six-speed-manual or four-speed automatic transmission, and a 3.23:1 final drive ratio.

Car Craft said the Firehawk had "the best performance-to-cost ratio of any car in its class," and advised readers "think of Firehawk as Corvette Lite—a third less costly, and it's more thrilling." The $24,244 car moved from 0–60 miles per hour in 4.9 seconds and did the quarter mile in 13.53 seconds at 103.5 miles per hour. SLP advertised that it was going to build only 250 Firehawks and it did. Even though low, that's 10 times as many as were made in 1992.

1993 Firebird Models and Base Prices

Model Number	Model Name	Dealer Cost	Suggested Retail Price
S87P	Firebird Coupe	$12,805	$13,995
V87P	Formula Coupe	$16,465	$17,995
V87P	Trans Am Coupe	$19,576	$21,395

Dealer destination charge (all models, effective Jan. 6, 1993):$490

Basic standard equipment: 3.4-liter fuel-injected V-6; power front disc brakes; five-speed manual transmission; and P215/60R16 black sidewall touring tires.

1993 Firebird Options and Accessories

UPC Code	Description	Retail Price
Option Packages		
Firebird		
1SA	Includes vehicle standard equipment only	n/c
1SB	Includes Option package 1SA plus (C60) air conditioning; (B35) front and rear floor mats; (B84) body side moldings; and AH3 4-way driver's side manual seat adjuster.	1,005.00
1SC	Includes 1SB plus (A31) power windows; (AU3) power door locks; (K34) cruise control; (DG7) left-hand power Sport mirror and right-hand convex power Sport mirror with blue glass.	1,836.00
1SD	Includes 1SC plus (AUO) remote keyless entry; (UX1) AM/FM ETR radio with auto reverse, cassette, five-band equalizer, and 10-speaker system; and leather appointment group.	2,471.00
Formula		
1SA	Includes vehicle standard equipment only	n/c
1SB	Includes 1SA plus (B35) front and rear floor mats; (B84) body side moldings; (A31) power windows; (AU3) power door locks; (K34) cruise control; (DG7) left-hand power Sport mirror and right-hand convex power Sport mirror with blue glass.	906.00
1SC	Includes 1SB plus (AUO) remote keyless entry; (UX1) AM/FM ETR radio with auto reverse, cassette, five-band equalizer, and 10-speaker system; and leather appointment group	1,541.00
Trans Am		
1SA	Includes vehicle standard equipment only	n/c
Other options		
Y83	Trans Am Option. Includes articulating custom seats; (G92) hi-performance axle and engine oil cooler; (B84) color-keyed body side moldings; front and rear floormats; (AM3) 4-way manual adjustable driver's seat; (A31) power windows; (K34) cruise control; (DG7) left-hand power Sport mirror and right-hand convex power Sport mirror with blue glass; (AUO) remote keyless entry; (UX1) AM/FM ETR radio with auto reverse, cassette, five-band equalizer, and 10-speaker system; (C49) electric rear window defogger; (OLC) P245/50ZRRED16 tires; leather appointment group; fog lamps; and specific Trans Am appearance package.	Std.
C60	Custom air conditioning. Included with 1SB, 1SC, 1SD on Firebird. Standard on other models.	895.00
C41	Non-air conditioning. Standard: Firebird. Not available: Trans Am. Available: 1SA on Formula.	(895.00)
G92	High-performance axle ratio and oil cooler. Required with QLC tires. Not available: Firebird. Standard: Trans Am	110.00
B84	Body color side moldings. Standard: Trans Am	60.00
NB2	Required California emissions	100.00
K34	Cruise control with resume speed. Standard: Trans Am	225.00
A90	Remote deck lid release. Requires 1SB on Firebird with LO3 V-8; Included in 1SC Formula and Trans Am option package. Standard in GTA.	60.00
C49	Electric rear window defogger. Standard: Trans Am.	170.00
CC1	Locking hatch roof. Removable with locks and stowage	895.00
DE4	Sunshades. Removable, for hatchback roof. Requires CC1. Not available with 1SA nor 1SB on Firebird, nor with 1SA on Formula.	25.00
VK3	Front license-plate bracket	n/c
B35	Rear floor mats. Standard: Trans Am. Available: 1SA on Firebird and Formula. Included in all other option packages.	15.00
DG7	Dual Sport power mirrors. Requires A31 windows and AU3 locks. Standard: Trans Am. Included in Firebird packages 1SC and 1SD; Formula packages 1SB and 1SC. Not available with 1SA packages.	96.00

UPC Code	Description	Retail Price
AU3	Power door locks (Available only and required with A31 power windows. Standard: Trans Am. Included in Firebird packages 1SC and 1SD and Formula packages 1SB and 1SC. Not available with 1SA on Firebird.	220.00
A31	Power windows. Available only and required with AU3 power door locks. Includes door map pockets. Standard with GTA. Other models:	290.00

Radio Equipment

UPC Code	Description	Retail Price
UX1	AM/FM ETR stereo with cassette, graphic equalizer, and clock. Consists of Delco AM stereo/FM ETR stereo radio and cassette with auto reverse and search-and-replay, and electronic seek-and-scan tuning. Integral 5-band graphic equalizer. Integral digital clock and dual front and rear co-axial speakers. Included in 1SD on Firebird and 1SC on Formula. Not available with 1SA on Firebird. Requires A31 windows and AU3 door locks.	500.00
U1A	Delco AM stereo/FM ETR stereo and CD player with electronic seek up/down tuning; next and previous CD controls; integral 5-band graphic equalizer; integral digital clock; and dual front and rear co-axial speakers. Not available with 1SA on Firebird. Requires A31 power windows and AU3 power door locks.	
	For Firebird with 1SB or 1SC and Formula with 1SA or 1SB.	726.00
	For Firebird with 1SC; Formula with 1SC; and Trans Am.	226.00
AUO	Remote Keyless Entry System	135.00

Seats

UPC Code	Description	Retail Price
AR9	Articulating bucket seats with leather seating surfaces. Not available with AC3 power seat. Firebird (available only w/1SD) or Formula (available only w/1SC)	780.00
AQ9	Articulating bucket seats with leather seating surfaces. Not available with AC3 power seat. Trans Am	475.00
AH3	4-way manual adjustable driver's seat. Not available: Formula. Standard: Trans Am. Included in Firebird 1SB, 1SC and 1SD packages	35.00
AC3	6-way power driver's seat. Not available on Trans Am or with AR9/AQ9 seating.	
	On Firebird with 1SA and Formula	305.00
	On Firebird with 1SB, 1SC or 1SD	270.00

Tires

UPC Code	Description	Retail Price
QMT	P235/55R16 black sidewall touring tires Standard: Formula. Not available: Trans Am. Optional: Firebird only	132.00
QLC	P245/50ZR16 black sidewall tires; and high-performance suspension. Not available: Firebird. Standard: Trans Am. Optional: Formula only	144.00

Miscellaneous

UPC Code	Description	Retail Price
MXO	Four-speed automatic transmission	595.00
WDV	New York State warranty enhancement	25.00
40P	White paint	n/c

1994

★	Firebird Convertible
★	Formula Coupe
★★	Formula Convertible
★★	Trans Am Coupe
★★★	Trans Am GT Coupe
★★★★	Trans Am GT Convertible
★★★(+1/2)	SLP Firehawk
★★★★	25th Anniversary Trans Am Coupe
★★★★★	25th Anniversary Trans Am Convertible

The release date for 1994 Pontiacs was September 21, 1993. The good-looking Firebirds introduced just months earlier were little changed, but with a full year to sell them, Pontiac predicted a 176 percent sales increase to 55,000 Firebirds. Actually, 51,523 were ultimately made.

"Firebird, the epitome of Pontiac sports-car-driving excitement for the '90s, becomes an even greater value for 1994, as all models carry over pricing from 1993," announced the company. "The price position increases Firebird's competitive edge, especially for the Formula, which provides, as standard equipment, dual airbags, anti-lock brakes, and 275 horsepower for under $18,000—a combination of safety, performance, and value unmatched by any competitor."

Advertising for the cars actually kicked off in the summer of '93, when Pontiac issued a challenge to other carmakers to match the Formula's features at even twice the price. "Pontiac believes these elements of the Firebird's message, particularly the introduction of state-of-the-art safety features to a world-class performance vehicle, represent a significant advance in the sports-car arena," said an official statement. "This will appeal not only to traditional Firebird buyers, but also to those buyers who have traditionally considered only imports."

New-for-1994 features included a Dark Aqua Metallic color; flood-lit interior door-lock switches; visor straps; a Delco 2001 Series radio family; a compact-disc player without equalizer; a 5.7-liter SFI V-8 with Mass-Airflow control system; four-speed electronically-controlled automatic transmission with driver-selectable automatic-transmission controls; updated six-speed manual transmission with a first-to-fourth gear "skip shift" feature; traction control (released at midyear for cars with a V-8 and automatic transmission only); and a two-component clearcoat exterior finish.

The year began with much the same offerings as in 1993. Pontiac called the Firebird "a new standard in sports cars." The base engine was again the L32 SFI 3.4-liter V-6 that produced 160 horsepower and 200 foot-pounds of torque. It was linked to a five-speed manual transmission or available four-speed automatic with driver-selectable "normal" and "second-gear-select" transmission modes.

The Formula was the first of three models to come standard with a sequential-port-fuel-injection 5.7-liter V-8 and six-speed manual transmission. This 275-horsepower LT1 V-8 was optionally available with the four-speed automatic. With a V-8, the driver-transmission-select switch offered "normal" and "performance" modes. Formulas also got wider P235/55R16 base tires; 16x8-inch Sport cast aluminum wheels; standard air conditioning; four-wheel disc brakes; and a Performance Ride & Handling suspension.

The '94 Firebird had a traditional twin-slot Pontiac grille, and Firebird lettering on the sides of the body. The convertible was officially introduced at the Chicago Auto Show in mid-February, but didn't get into showrooms until months later.

The Formula had the same grille as the base Firebird, but lacked any lettering on the side of the body. It had body-color Sport mirrors standard, and neutral-density taillamp lenses at the rear. A 5.7-liter V-8 was the only engine used in the Formula.

Again the Formula had no body-side lettering, but "Formula V8" identification appeared on the left-hand headlamp cover and right-hand side of the rear bumper. The Formula retained body-color Sport mirrors, and smoothly-contoured taillamps with neutral-density lenses (also used on Trans Ams and Trans Am GTs).

Pontiac advertised that the Trans Am had "the power to change minds." The company suggested that it represented a mature approach to power, because its performance capabilities were "properly managed." It was technically much the same as the Formula, except that it used speed-rated, high-performance P245/50ZR16 Goodyear Eagle GSC tires.

Externally, the "TA" gained a new, uniquely-styled front end without Pontiac's traditional twin-air-slot grille. Instead, the nose had a blunter look, broken by round, sunken running lights on each side of center. The Trans Am name appeared on the body sides, and between the taillamps (along with a "screaming-chicken" emblem).

A new model was added to the top of the Trans Am line up. The Trans Am GT featured body-colored body-side moldings as standard equipment, plus a special GT spoiler (it looked more like an air foil than the other spoiler). Also considered standard on this car was an electric rear-window defogger; Remote Keyless Entry system; rear carpet mats; Delco 2001 Series sound system; four-way manual driver's seat; and leather-wrapped steering wheel, shift knob, and parking-brake handle.

On January 27, 1994, Pontiac announced that it planned to release a special model to honor the

The Trans Am had a unique front-end design, and more prominent lower body skirting. The name Trans Am appeared on the sides of the body. This is a Trans Am GT, as the regular Trans Am did not come as a convertible.

The special 16x8-inch Sport Cast aluminum wheels on cars with the 25th Trans Am option were finished in white. This car had a top speed of 152 miles per hour, compared to the '69 Trans Am's 135-mile-per-hour estimated top speed. The anniversary option package retailed for $995.

silver anniversary of the Trans Am. This car was officially described as "the 25th Anniversary Edition Trans Am," although it was also called simply "25th Trans Am." This is what actually appears on the body sides the wheel caps, and the seats. It started production later in January, and appeared in Pontiac showrooms that spring. It was the fourth anniversary-model Trans Am—the others having been produced in 1979, 1984, and 1989.

The $995 price of the 25th Anniversary package included bright white exterior finish; a bright blue centerline stripe; anniversary logos and door badges; lightweight 16-inch aluminum wheels painted bright white; and white Prado leather seating surfaces with blue embroidery. The seat backs had "25th Trans Am" embroidered into them in blue, and similar embroidery appeared

Sales catalog art showed the 1994 Trans Am GT interior with available leather seating surfaces. A new-generation Delco 2001 Series stereo system was available for the first time this year. This car has the optional electronic-controlled automatic transmission.

Blue "25th Trans Am" embroidery adorned the seat backs and door panels of the limited-edition anniversary-model Trans Am. This car features the six-speed manual transmission, which was standard with the 5.7-liter (350-ci) V-8.

on the door inserts. Buyers of the cars received a special 25th Anniversary portfolio.

"The Trans Am symbolizes Pontiac driving excitement at its highest level, and it's for that reason we are introducing an anniversary edition on its 25th birthday," said Pontiac General Manager John G. Middlebrook. "Pontiac will only build a limited number of the highly-contented 25th Anniversary Trans Ams."

A really neat touch was the news that a very limited number of soon-to-be-released Trans Am GT convertibles would be optioned as anniversary models, in honor of the eight ragtops made in the Trans Am's first year. In addition to a special press kit for the 25th Trans Am, Pontiac released a four-page sales brochure headlined "A blast from . . . the future." It showed both the '69 and '94 Trans Ams, with specifications charts for both cars.

The year wasn't over yet for Firebird fans. On February 4, 1994, PMD announced that it was introducing the 1994 Firebird convertible at the Chicago Auto Show. A press kit indicated that the ragtop would be available "on Firebird, Formula, and Trans Am models," although only the Trans AM GT came in the open-air-driving style. The release date for this body style was also in the spring.

All convertibles featured a flush-folding power top that could be stored beneath an easily-assembled three-piece tonneau cover. They included a glass rear window and electric rear-window defogger. "These Firebird convertibles weren't an afterthought," revealed Pontiac Chief Engineer Byron Warner. "Before the designers began sketching the cars, we knew we wanted convertibles, so structural integrity was designed into the vehicle from the start."

Warner said that 13 structural parts were added to the Firebird, and front cross member structural stiffness was increased. Structural adhesive was also utilized on 15 key areas, such as the cowl, pillar, and rocker reinforcements. Convertibles were built on the Ste. Therese assembly line, to ensure high quality. Warner said that ragtop drivers "will think they're driving one of our coupes when the top is up."

For collector's, the car of this vintage to have is the very-rare 25th Trans Am convertible. The coupe ties for second-most-desirable model with the Firehawk from SLP Engineering, of which 500 were made this year. (See 1993 section.) Beyond that, any convertible is probably worth investing in, and the Trans Am GT version would still be a safe bet as a future investment.

1994 Firebird Models and Base Prices

Model Number	Model Name	Dealer Cost	Suggested Retail Price
S87R	Firebird Coupe	$13,130	$14,349
S67R	Firebird Convertible	$19,608	$21,429
V87R	Formula Coupe	$16,982	$18,559
V67R	Formula Convertible	$22,499	$24,589
V87R, Y83	Trans Am Coupe	$18,592	$20,319
V87R, Y83	Trans Am GT Coupe	$19,965	$21,819
V67R, Y83	Trans Am GT Convertible	$24,512	$26,789

Dealer destination charge (all models effective May 16, 1994): $490

Basic standard equipment: 3.4-liter fuel-injected V-6; power front disc brakes; five-speed manual transmission; and P215/60R16 black sidewall touring tires.

1994 Firebird Options and Accessories

UPC Code	Description	Retail Price ($)
Option packages		
Firebird Coupe		
1SA	Includes vehicle standard equipment only	n/c
1SB	Includes Option package 1SA plus (C60) air conditioning; (B35) rear floormats; (B84) body side moldings; and (AH3) 4-way driver's side manual seat adjuster.	1,005.00
1SC	Includes 1SB plus (A31) power windows; (AU3) power door locks; (K34) cruise control; (DG7) left-hand power Sport mirror and right-hand convex power Sport mirror with blue glass; (AUO) remote keyless entry; (UX1) AM/FM ETR radio with auto reverse, cassette, five-band equalizer, and 10-speaker system; and leather appointment group.	2,421.00
Formula Coupe		
1SA	Includes vehicle standard equipment only	n/c
1SB	Includes 1SA plus (B35) front and rear floormats; (B84) body side moldings; (A31) power windows; (AU3) power door locks; (K34) cruise control; (DG7) left-hand power Sport mirror and right-hand convex power Sport mirror with blue glass.	906.00
1SC	Includes 1SB plus (AUO) remote keyless entry; (UX1) AM/FM ETR radio with auto reverse, cassette, five-band equalizer, and 10-speaker system; and leather appointment group	1,491.00
Trans Am Coupe		
1SA	Includes vehicle standard equipment only	n/c
Firebird Convertible		
1SA	Includes vehicle standard equipment only	n/c
1SB	Includes (AUO) remote keyless entry; AM/FM ETR radio with auto reverse, cassette, five-band equalizer, and 6-speaker system; and leather appointment group	485.00
Formula Convertible		
1SA	Includes vehicle standard equipment only	n/c
1SB	Includes (AUO) remote keyless entry; AM/FM ETR radio with auto reverse, cassette, five-band equalizer, and 6-speaker system; and leather appointment group.	485.00
Trans Am GT Convertible		
1SA	Includes vehicle standard equipment only	n/c
Other options		
Y83	Trans Am GT option. Includes articulating custom seats; (G92) high-performance axle and engine oil cooler; (B84) color-keyed body side moldings; front and rear floormats; (AM3) 4-way manual adjustable driver's seat; (A31) power windows; (K34) cruise control; (DG7) left-hand power Sport mirror and right-hand convex power Sport mirror with blue glass. (AUO) remote keyless entry; (UX1) AM/FM ETR radio with auto reverse, cassette, five-band equalizer, and 10-speaker system; (C49) electric rear window defogger; (OLC) P245/50ZRRED16 tires; leather appointment group; fog lamps; and specific Trans Am appearance package.	Std.
C60	Custom air conditioning. Included with 1SB, 1SC, 1SD on Firebird. Standard on other models.	895.00
C41	Non-air conditioning. Standard Firebird. Not available on Trans Am. Available with 1SA on Formula.	(895.00)
B84	Body color side moldings. Standard on Trans Am.	60.00
K34	Cruise control with resume speed. Standard on Trans Am.	225.00
C49	Electric rear window defogger. Standard on Trans Am.	170.00
CC1	Locking hatch roof—Removable, with locks and stowage.	895.00
DE4	Sunshades—removable, for hatchback roof. Requires CC1. Not available with 1SA or 1SB on Firebird, nor with 1SA on Formula.	25.00
VK3	Front license-plate bracket	n/c
B35	Rear floormats. Standard on Trans Am and available with 1SA on Firebird and Formula. Included in all other option packages.	15.00

UPC Code	Description	Retail Price ($)
DG7	Dual Sport power mirrors. Standard Trans Am. Included in Firebird packages 1SC and 1SD; Formula packages 1SB and 1SC. Not available with 1SA packages. Requires A31 windows and AU3 locks.	96.00
AU3	Power door locks. Available only and required with A31 power windows. Standard on Trans Am. Included in Firebird packages 1SC and 1SD and Formula packages 1SB and 1SC. Not available with 1SA on Firebird.	220.00
A31	Power windows. Available only and required with AU3 power door locks. Includes door map pockets. Standard with GTA. Other models:	290.00
GU5	Performance rear axle. Not available on Firebird, Trans Am or Trans Am GT.	175.00

Radio Equipment

UPC Code	Description	Retail Price ($)
U1C	Delco AM stereo/FM ETR stereo and CD player with electronic seek up/down tuning; next and previous CD controls; integral 5-band graphic equalizer; integral digital clock; and dual front and rear co-axial. speakers. Not available with 1SA on Firebird. Requires A31 power windows and AU3 power door locks. Firebird with 1SC; Formula with 1SC; and Trans Am:	226.00
UT6	ETR AM/FM stereo with auto reverse, cassette, and graphic equalizer. Standard Trans Am and Trans Am GT.	
	On other coupes:	450.00
	On other convertibles:	350.00
UT3	ETR AM/FM stereo with CD; and graphic equalizer. Standard Trans Am and Trans Am GT.	
	On other coupes:	676.00
	On other convertibles:	570.00
	On cars with 1SB, 1SC:	226.00
AUO	Remote Keyless Entry System	135.00

Seats

UPC Code	Description	Retail Price ($)
AH3	4-Way manual adjustable driver's seat. Not available: Formula. Standard: Trans Am. Included in Firebird 1SB, 1SC and 1SD packages.	35.00
AC3	Six-Way power driver's seat. Not available on Trans Am or with A-9 seating.	
	Firebird with 1SA, Formula	305.00
	Firebird with 1SB, 1SC, 1SD	270.00

Tires

UPC Code	Description	Retail Price ($)
QMT	P235/55R16 black sidewall touring tires. Standard: Formula. Not available: Trans Am. Optional on Firebird only:	132.00
QLC	P245/50ZR16 black sidewall tires; and high-performance suspension. Not available: Firebird. Standard: Trans Am. Optional on Formula only:	225.00

Miscellaneous

UPC Code	Description	Retail Price ($)
MXO	Four-speed automatic transmission	775.00
WDV	New York State warranty enhancement	25.00

1995

★	Firebird Convertible
★	Formula Coupe
★★	Formula Convertible
★★★	Trans Am Coupe
★★★(+1/2)	Trans Am Convertible
★★★★(+1/2)	Firehawk Coupe
★★★★★(+1/2)	Firehawk Convertible

Pontiac's lone rear-wheel-drive car, the Firebird, continued to be built in a Canadian factory. It was offered, for 1995, in Firebird, Formula, and Trans Am models. Each came as a coupe or a convertible. SLP Engineering also built limited-edition, high-performance versions of both body styles that could be ordered through Pontiac dealers.

"Firebird is styled to be noticed, equipped for outstanding performance and handling, and priced to be accessible," said Pontiac's general manager John Middlebrook, at the Chicago Auto Show. "A lot of people are taking notice, because for the 1994 model year, Firebird sales were up 143 percent over the previous year. We look for the exciting lineup of 1995 Firebirds to continue a hot sales pace."

Traction Control was now offered for Firebird V-8 models equipped with either of two trans-missions, four-speed automatic or six-speed manual. (It was not offered for the base Firebird V-6, which used a five-speed manual gear box). Three new exterior colors were introduced: blue-green chameleon, medium dark purple metallic, and bright silver metallic. Inside, a new bright red leather interior option was offered for all models, while a bright white leather interior was an extra that was limited to convertibles.

Other new features included 16-inch, five-spoke wheels for V-8 models; all-weather speed-rated P245/50Z16 tires; a power antenna; and a four-spoke Sport steering wheel. Firebird models with optional up-level sound systems were factory pre-wired to receive a Delco 12-disk CD changer, which was available as a new dealer-installed option. All 1995 Firebirds also had maintenance-free lower control-arm ball joints and "lubed-for-life" front-end components. Because Pontiac wheels were designed to show the wheel-and-brake mechanical parts, all Firebird models got a special coating to keep the brake rotors looking like new. Additionally, a new 16-inch, five-spoke wheel was available in aluminum for Formulas, and chrome for Trans Ams.

By 1995, even the base Firebird was a heavily-contented car. It featured soft fascia-type bumpers and composite body panels on the doors, front fenders, roof, rear deck lid, and rear spoiler. Electrically-operated concealed quartz-halogen headlamps were standard, along with one-piece, multi-color-lens taillamps and a center high-mounted stop lamp. Body-color body-side moldings were on convertibles. Dual, power remote-control Sport mirrors with blue glass were fitted. The cars were finished with a waterborne base coat and two-component clearcoat system. A

For 1995, the base Firebird convertible came with a long list of standard equipment. Dual air bags and four-wheel anti-lock brakes were standard equipment.

"Now you're playing with fire" was Pontiac's advertising slogan for the 1995 Formula coupe. It came with the LT1 5.7-liter (350-ci) V-8 which produced 275 horsepower at 5,000 rpm and 325 foot-pounds of torque at 2,400 rpm.

The Firebird Formula convertible listed for more than $25,000 in 1995. Pontiac built around 49,000 of all Firebird models this season. The Formula is considered to have a high-performance image, as opposed to the Trans Am's luxury-performance image.

The 1995 Trans Am coupe and convertible shared their drivetrain with Formula models. This model weighed 3,445 pounds, had a 101.1-inch wheelbase and a 197-inch overall length. It was rated for 17 miles per gallon around town and 25 miles per gallon on the highway.

black, fixed-mast antenna was mounted at the right rear of the body. Solar-Ray tinted glass was standard.

Inside, the Firebird had reclining two-way manual bucket seats with standard Matrix cloth trim; full-floor cut-pile carpeting; driver and passenger air bags; an analog speedometer; full gauges; a high-tech Delco sound system; Remote Keyless Entry; and PASS-Key II theft-deterrent system. Convertibles included cruise control, more lamps, and power windows.

Tires were P215R/60R16 steel-belted radials. They came on machine-faced, cast aluminum wheels with gray ports. A compact spare was provided. Controlled-cycle wipers were standard, and convertibles also included cruise control and an electric rear-window defogger. Gas-charged de Carbon shocks were featured, along with a Delco Freedom II battery and driver-selectable transmission controls.

The RPO L32 3.4-liter V-6 with Computer Command Control produced 160 horsepower at 4,600 rpm and 200 foot-pounds of torque at 3,600 rpm. It came standard with the five-speed manual transmission. An electronically-controlled four-speed automatic was optional with the same engine. A 3.23:1 ratio rear axle was used. The suspension was of the short-and-long arm type, with front-and-rear stabilizers. Power rack-and-pinion steering was included, too.

An optional 3.8-liter V-6 for base Firebirds only was offered this season, as a $350 option. This L36 engine, also known as the "3800 Series II," was popular in the Bonneville, and was a midyear addition to the Firebird's option list announced at the Chicago Auto Show in February. It featured SFI and delivered 200 horsepower and 225 foot-pounds of torque. The 3800 Series II was mated to Pontiac's powerful 4L60-E automatic transmission.

In addition to (or in place of) base Firebird features, Formula models included smooth-contour neutral-density taillamps; P235/55R16 SBR touring tires; bright silver 16x8-inch Sport Cast aluminum wheels; air conditioning; a 125-amp alternator; four-wheel power disc brakes; a 5.7-liter V-8; a six-speed manual transmission; and a performance suspension. With automatic transmission, you got driver-selectable transmission controls with "normal" and "performance" modes.

The top-of-the-line Trans Am had a number of added or replacement standard features over those included with Formulas, such as: fog lamps; body-color, body-side moldings; P245/50ZR16 speed-rated, all-weather tires; cruise control; automatic power door locks; and a "juicier" Delco sound system. Rear seat courtesy lamps and a trunk lamp were deleted, however.

Promoted as "an irresistible force that moves

A six-speed Hurst shifter was new for the high-performance 1995 Firehawk, so SLP Engineering photographed a Firehawk coupe with a '65 Pontiac Catalina 2 + 2, which also came with a Hurst shifter.

you," the 1995 Firehawk version of the Firebird offered 300- and 315-horsepower versions, plus all-new features like optional chrome wheels. In addition, Ed Hamburger and his SLP engineers added a Hurst six-speed shifter. About 750 Firehawks were sold this year.

Twin air scoops on the Firehawk's snout were part of a special air-induction system that added 25–35 horsepower. The suspension was upgraded from Formula specifications to increase handling limits. Coupes benefited from bigger, better-handling Firestone 275/40-17 tires and 17-inch wheels. (Firehawk convertibles, however, came only with Firestone 16-inch tires and closed-lug 16-inch alloy wheels.) Other features included megaphone-style polished stainless-steel tailpipe tips, special exterior graphics, and a numbered dash plaque. A sport suspension and performance exhaust system were optional.

A six-speed Firehawk coupe with the performance exhaust system went from 0–60 miles per hour in 4.9 seconds and did the quarter mile in 13.5 seconds at 103.5 miles per hour. Top speed was 160 miles per hour. The Firehawk was available as an SLP alteration on '95 Formulas, which buyers ordered through Pontiac dealers. They came with a three-year, 36,000-mile limited warranty. Base price for the alteration was $6,495, an increase of $500 over 1993–1994.

1995 Firebird Models and Base Prices

Model Number	Model Name	Dealer Cost	Suggested Retail Price
S87S	Firebird Coupe	$13,820	$15,104
S67S	Firebird Convertible	$20,160	$22,039
V87S	Formula Coupe	$17,700	$19,344
V67S	Formula Convertible	$23,085	$25,229
V87S, Y82	Trans Am Coupe	$19,383	$21,184
V67S, Y82	Trans Am Convertible	$24,294	$27,239

Dealer destination charge (all models effective Feb. 10, 1995): $500

Basic standard equipment: 3.4-liter fuel-injected V-6; power front disc brakes; five-speed manual transmission; and P215/60R16 black sidewall touring tires.

1995 Firebird Options and Accessories

UPC Code	Description	Retail Price ($)
Engines		
L32	3.4-liter SFI V-6. Standard: Firebird. Not available: Formula nor Trans Am	Std.
LT1	5.7-liter SFI V-8. Standard: Formula, Trans Am. Not available: Firebird	Std.
L36	3.8-liter SFI V-8. Requires MXO transmission, C60 air conditioning.	
	Not available: Formula, Trans Am	350.00
Emission Systems		
FE9	Federal	n/c
YF5	California	100.00
NG1	Massachusetts	100.00
Tires		
QPE	P215/60R16 Blackwall. STL touring. Standard on Firebird. Not available: Formula, Trans Am	n/c
QCB	P235/55R16 Blackwall. Touring. Standard on Formula. Not available: Trans Am	132.00
QFZ	P245/50ZR16 Blackwall. All-weather performance. Includes 150-mile-per-hour speedometer.	
	Requires NW9 Traction Control. Not available: base Firebird	225.00
QLC	P245/50ZR16 Blackwall. High performance. Includes 150-mile-per-hour speedometer.	
	Not available: base Firebird, nor with NW9 Traction Control	
	On Formula	225.00
	On Trans Am	225.00
Interiors		
B9	Bucket seats with Matrix cloth trim. Firebird and Trans Am	Std.
AQ9	Articulating bucket seats with Matrix cloth trim. Trans Am only	330.00
AQ9	Articulating bucket seats with Prado leather trim	
	Requires packages 1SC and UT6 or UP3 radio	
	Formula coupes	804.00
	Trans Am coupes	829.00
	Convertibles (requires package 1SB)	829.00
Sound Systems		
UN6	ETR AM/FM stereo with auto reverse cassette. Standard all models.	Std.
U1C	ETR AM/FM stereo with CD player. Not available in Trans Am.	100.00
UT6	ETR AM/FM stereo with auto-reverse cassette; graphic equalizer; and power antenna.	
	Firebird and Formula coupes	473.00
	Firebird and Formula convertibles	373.00
	Trans Am coupe	398.00
UT3	ETR AM/FM stereo with compact-disc player and graphic equalizer	
	In Firebird and Formula coupe without 1SC	573.00
	In Firebird and Formula coupe with 1SC	100.00
	In Trans Am coupe	498.00
	In Firebird and Formula convertible without 1SB	473.00
	In Firebird and Formula convertible with 1SB	100.00
	In Trans Am convertible	100.00
NW9	Traction Control system.	450.00
Transmissions		
MM5	Five-speed manual. Standard: Firebird. Not available on other models.	Std.
MN6	Six-speed manual. Standard: Formula, Trans Am. Not available: Firebird.	Std.
MXO	Four-speed automatic	775.00
Option Groups		
	Firebird Coupe	
1SA	Includes vehicle standard equipment only	n/c
1SB	Includes air conditioning; rear floor mats; body-side moldings; and 4-way adjustable seat.	1,005.00

UPC Code	Description	Retail Price ($)
1SC	Includes 1SB; power windows and door locks; cruise control; power mirrors; remote keyless entry system; graphic equalizer with steering wheel controls; 10 speakers; leather steering wheel; power antenna; and electric rear-window defogger.	2,614.00
	Formula Coupe	
1SA	Includes vehicle standard equipment only	n/c
1SB	Includes rear floor mats; body-side moldings; power door locks and power windows; cruise control; power mirrors; and electric rear-window defogger.	1,076.00
1SC	Includes 1SB; remote keyless entry system; graphic equalizer with steering wheel controls; 10 speakers; leather steering wheel; and power antenna.	1,684.00
	Trans Am Coupe	
1SA	Includes vehicle standard equipment only	n/c
	Firebird Convertible	
1SA	Includes vehicle standard equipment only	n/c
1SB	Remote keyless entry system; graphic equalizer; 6-speaker sound system with steering wheel controls; leather appointment group; and power antenna.	508.00
	Formula Convertible	
1SA	Includes vehicle standard equipment only	n/c
1SB	Remote keyless entry system; graphic equalizer; 6-speaker sound system with steering wheel controls; leather appointment group; and power antenna.	508.00
	Trans Am Convertible	
1SA	Includes vehicle standard equipment only	n/c

Other options

UPC Code	Description	Retail Price ($)
Y82	Trans Am option. Not available on Firebird.	Std.
C60	Air conditioning. Standard: Formula, Trans Am	895.00
C41	Non-air-conditioned coupes. Standard: Firebird	(895.00)
B84	Body-color body-side moldings, for coupes. Standard: Trans Am	60.00
K34	Cruise control with resume speed on coupes. Standard: Trans Am	225.00
C49	Rear-window defogger on coupes. Standard: Trans Am	170.00

UPC Code	Description	Retail Price ($)
CC1	Removable hatch roof with locks and stowage. Coupes only. Not available with 1SA on Firebird or Formula.	970.00
DE4	Hatch roof sunshades. Requires CC1 hatch roof option. Not available with 1SA on Firebirds or Formulas.	25.00
VK3	Front license plate bracket	n/c
DG7	Left- and right-hand power mirrors with blue glass, on coupes. Standard: Trans Am	96.00
AU3	Power door locks, on coupes. Standard: Trans Am	220.00
A31	Power windows, on coupes. Standard: Trans Am	290.00
T43	Up-level rear spoiler. Trans Am coupe only.	395.00
GU5	Performance rear axle. Not available on Firebird.	175.00
AUO	Remote Keyless Entry. Standard: Trans Am convertible	135.00
AH3	4-way manual adjustable driver's side seat, coupes. Not available: Formula	36.00
AG1	6-Way power adjustable driver's seat. Not available with articulating bucket seats nor leather trim groups.	
	Firebird coupes with 1SA:	305.00
	Firebird coupes with 1SB or 1SC:	270.00
	Formula, Trans Am coupes:	305.00
	Convertibles:	270.00

1996

The Firebird roared into 1996 with more excitement and more power. There were new performances packages for both V-6 and V-8 models. Factory offerings included base Firebird, Formula, and Trans Am models. All came in coupe and convertible body styles. Formula and Trans Am coupes had a new WS6 Pontiac Ram Air performance-and-handling package that was instantly desirable, and collectible. The Ram Air cars were promoted as if they were separate models.

Though it actually arrived in mid-1995, the powerful 3800 Series II V-6 was pushed as a new-for-'96 feature. Rated at 200 horsepower at 5,200 rpm, the 3.8-liter V-6 developed 225 foot-pounds of torque at 4,000 rpm. It had a higher 6,000-rpm redline, and required spark-plug changes only once every 100,000 miles.

Newly available was a 3800 performance package for base coupes and ragtops. It included four-wheel disc brakes; a limited-slip differential; a dual-outlet exhaust system; P235/55R16 tires; an up-level steering gear ratio of 14.4:1 (compared to the stock 16.9:1); and a specific axle ratio (3.42:1 with automatic and 3.08:1 with manual transmission).

Horsepower of the LT1 V-8 was increased to 285, thanks to exhaust-system improvements. With the WS6 option, this leaped to 305 horsepower at 5,400 rpm, while torque was 335 foot-pounds at 3,200 rpm. WS6 also included a special twin-port hood scoop with Ram Air logos below each "nostril;" 17x9-inch cast aluminum five-spoke wheels; P275/40ZR17 tires; a dual-pipe catalytic converter system; aluminum exhaust tips; and specific suspension tuning.

The suspension upgrade included 32-mm

The 1996 Firebird coupe. Pontiac said, "If it were in your blood, it'd be adrenaline; If it were more advanced, it'd be from NASA; If it came from Italy, it'd be way too expensive!"

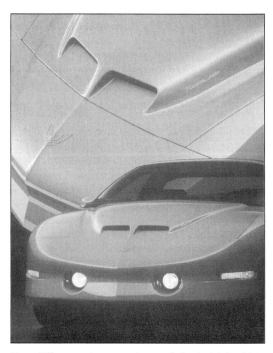

Two different close-up views of the hairy-looking WS6 twin-nostril hood. Pontiac said the car "bridged the gap between the raw power of the Muscle Car Era and today's sophisticated performance and safety technology."

The Firebird convertible came with a flush-folding power top. Thirteen specific structural parts were added to Firebird ragtops, and the stiffness of the front crossmember was increased.

The 1996 Formula with the WS6 option, like this T-top coupe, came equipped with fat, 17-inch (diameter) Goodyear Eagle GS-C tires. This was an ultra high-performance tire featuring an asymmetrical, directional tread design.

front and 19-mm rear sway bars; higher-rate front-and-rear springs; unique shock valving; stiffer transmission mounts; and stiffer Panhard-bar bushings. Both the V-6 and V-8s were OBD II compliant (meaning that they met stage II on-board-diagnostic system regulations to assure proper emission controls). OBD II was designed to monitor the EGR valve, oxygen sensor, and crankshaft- and camshaft-position sensors to detect misfiring.

Also new for the year was a theft-deterrent system with an available personal security key fob; a new taupe interior color; a new front seat belt buckle for improved child lockability; the availability of steering-wheel-mounted radio controls for all but the W51 sound system; the availability

Here's a Firebird Formula Coupe with no T-tops, but with the WS6 package. WS6 front springs were increased from the normal 51 Newton-meters stiffness to 63 Newton-meters. Rear-spring stiffness was variable from 23–30 Newton-meters on the WS6, compared to 23 Newton-meters on a normal Firebird V-8.

A WS6-equipped Trans Am. On the TA, the twin-snout hood was combined with the twin-foglamp front fascia. The WS6 Performance & Handling package could not be ordered for any '96 convertible. It came only on coupes.

The '96 Trans Am convertible was the high-priced Firebird this season. Structural adhesives were again used on 15 key areas of every soft-top body, including the cowl, windshield pillar, and rocker reinforcements.

Another Trans Am coupe with WS6. PMD described this as an "open-snouted, fire-breathing dragon made for serious driving enthusiasts who want the response of an I-Ain't-Kiddin' 305-hp V-8 when they press on the accelerator."

The "business department" of the WS6-equipped Formula or Trans Am was under the hood. The 5.7-liter engine featured sequential-port fuel injection, and an Opti-Spark ignition system.

of a power antenna on all Firebirds; a new coaxial four-speaker system option; a new red-orange metallic body color; and zinc brake-corrosion protection on all models.

In 1996, Pontiac learned that the Firebird had won the *Inside Automotives* magazine award for the "interior of the year" in the sports-car category, as well as Maryland Public TV's "Motor-Week" 1995 Driver's Choice Award for "Best Convertible."

Hot and fast
It's just the way we do things.

PONTIAC

FIREBIRD

Just watch us cook.

McDONALD'S RACING TEAM

Jim Yates

© 1996 McDonald's Corporation

Jim Yates drove the McDonald's/Joe Gibbs Racing National Hot Rod Association (NHRA) Pro Stock Firebird to a second place in the point standings in 1995, and was planning to be back at dragstrips across the country in 1996.

1996 Firebird Models and Base Prices

Model Number	Model Name	Dealer Cost	Suggested Retail Price
S87S	Firebird Coupe	$14,287	$15,614
S67S	Firebird Convertible	$20,536	$22,444
V87S	Formula Coupe	$17,810	$19,464
V67S	Formula Convertible	$23,135	$25,284
V87S, Y82	Trans Am Coupe	$19,594	$21,414
V67S, Y82	Trans Am Convertible	$25,038	$27,364

Dealer destination charge (all models effective Sept. 22, 1995): $505

Basic standard equipment: 3.8-liter fuel-injected V-6; power front disc brakes; five-speed manual transmission; and P215/60R16 black sidewall touring tires.

1996 Firebird Options and Accessories

UPC Code	Description	Retail Price ($)
Emission Systems		
FE9	Federal	n/c
YF5	California	n/c
NG1	Massachusetts	n/c
NB8	Override, Calif./N.Y./Mass. Requires FE9	n/c
n/c7	Override, Federal. Requires YF5 or NG1	n/c
Tires		
QCB	P235/55R16 Blackwall. Touring. Standard Formula. Not available on Trans Am.	132.00
QFZ	P245/50ZR16 Blackwall. All-weather performance. Includes 150-miles-per-hour speedometer. Requires (NW9) Traction Control. Not available on base Firebird.	225.00
QLC	P245/50ZR16 Blackwall. High performance. Includes 150-miles per hour speedometer. Not avail.: Firebird nor with (NW9) Traction Control	
	Formula:	225.00
	Trans Am:	225.00
QFK	P275/40ZR17 speed-rated tires. Available with WS6 only Formula, Trans Am coupes.	n/c
Interiors		
AR9	Articulating bucket seats with Prado leather trim. Includes adjustable lumbar and thigh support (AR9) Coupes only. Requires option package 1SC and W54 or W55 radio upgrade. Convertible requires 1SB and W59 or W73 radio upgrade.	
	Formula coupe	804.00
	Convertible	829.00
JAQ9	Includes AR9, plus articulating headrests	
	Trans Am coupe with Metrix cloth trim	330.00
	Trans Am coupe with Prado leather trim (req's. W54 or W55 radio)	829.00
Option Groups		
	Firebird Coupe	
1SA	Includes vehicle standard equipment only	n/c
1SB	Includes air conditioning; rear floormats; body-side moldings; 4-way adjustable seat; ETR AM/FM cassette with auto reverse; graphic equalizer, clock, seek up/down; and four speaker coaxial sound system.	1,078.00
1SC	Includes 1SB; power windows and door locks; cruise control; power mirrors; remote keyless entry system; graphic equalizer with steering wheel controls; power antenna; electric rear-window defogger; and leather appointment group.	2,499.00
	Formula Coupe	
1SA	Includes vehicle standard equipment only	n/c
1SB	Includes rear floormats; body-side moldings; 4-way seat adjuster; ETR AM/FM stereo with auto reverse cassette; graphic equalizer; clock; seek up/down; 4-speaker coaxial sound system; power windows; power locks; cruise control; power mirrors; electric rear window defogger.	1,184.00
1SC	Includes 1SB plus remote keyless entry; power antenna; steering wheel controls; with leather appointment group.	1,604.00
	Trans Am Coupe and Convertible	
1SA	Includes vehicle standard equipment only	n/c
Other options		
Y82	Trans Am coupe option. Not available on Firebird.	Std.
Y84	Trans Am convertible option. Not available on Firebird.	Std.

UPC Code	Description	Retail Price ($)
Y87	3800 performance package. Includes limited-slip differential; 4-wheel disk brakes; up-level steering; dual outlet exhaust; QXB tires; five-spoke aluminum wheels. Also includes 3.42:1 rear axle ratio if ordered with MXO transmission. Available on Firebird coupe and convertible only. Not available with 1SA on coupe.	535.00
WS6	Ram Air performance and handling package. Includes Ram Air induction system; functional air scoops; five-spoke 17-inch aluminum wheels; P275/40ZR17 speed-rated tires; and dual, oval exhaust outlets. Available on Formula and Trans Am coupes only.	2,995.00
B84	Body-colored body-side moldings. Standard on Trans Am.	60.00
UA6	Content theft alarm. Requires AUO keyless entry. Not available with 1SA on Firebird coupe.	90.00
K34	Cruise control with resume speed on coupes. Standard: Trans Am	225.00
C49	Rear-window defogger on coupes. Standard: Trans Am	170.00
CC1	Removable hatch roof with locks and stowage. Coupes only. Not available with 1SA on Firebird or Formula.	970.00
DE4	Hatch roof sunshades. Requires CC1 hatch roof option. Not available with 1SA on Firebirds or Formulas.	25.00
VK3	Front license plate bracket	n/c
B35	Front and rear carpet mats. Standard: Trans Am coupe and convertible, and included in all 1SB and 1SC packages.	15.00
DG7	Left- and right-hand power mirrors with blue glass on coupes. Standard: Trans Am	96.00
U75	Power antenna. Requires optional radio upgrade if ordered on Firebird or Formula coupe. Standard: Trans Am convertible	85.00
AU3	Power door locks, on coupes. Standard: Trans Am coupe, all convertibles	220.00
A31	Power windows, on coupes. Standard: Trans Am	290.00

Sound Systems

UPC Code	Description	Retail Price ($)
W52	ETR AM/FM stereo with auto reverse cassette; graphic equalizer; clock; seek up/down; remote CD prewire and four speakers coaxial system; Standard: Trans Am. Available on coupes only. Included in 1SB and 1SC.	73.00
W53	ETR AM/FM stereo with CD player; graphic equalizer; clock; seek up/down; and four-speaker coaxial sound system. Available in coupes only.	
	Firebird, Formula with 1SA:	173.00
	Firebird, Formula with 1SB or 1SC; Trans Am:	100.00
W54	ETR AM/FM stereo with auto reverse cassette; graphic equalizer; clock; seek up/down; remote CD prewiring; and 10-speaker sound system. Available in coupes only. Not available with 1SA on Firebird or Formula; Requires AU3 power locks and A31 power windows.	115.00
W55	ETR AM/FM stereo with CD player; graphic equalizer; clock; seek up/down; 10-speaker sound system. Available in coupes only. Not available with 1SA on Firebird or Formula. Requires AU3 power locks and A31 power windows.	215.00
W58	ETR AM/FM stereo with CD player; graphic equalizer; clock; seek up/down; 4-speaker coaxial sound system. Available in Firebird or Formula convertibles with 1SA:	100.00
W59	ETR AM/FM stereo with auto reverse cassette; graphic equalizer; clock; seek up/down; remote CD prewiring; and 6-speaker sound system; Available in convertibles only. Standard: Trans Am. Included in 1SB.	50.00
W73	ETR AM/FM stereo with CD player; graphic equalizer; clock; seek up/down; 6-speaker sound system; Available in convertibles only.	
	Firebird and Formula with 1SA:	150.00
	Firebird and Formula with 1SB or Trans Am:	100.00
US1	Remote 12-disc CD changer, trunk mounted in coupes. Requires W52 or W54 radio upgrade.	595.00
GU5	Performance rear axle. Not available on Firebird.	175.00

UPC Code	Description	Retail Price ($)
AUO	Remote keyless entry. Standard: Trans Am convertible.	135.00
AH3	4-way manual adjustable driver's side seat, coupes. Not available in Formula.	36.00
AG1	6-way power adjustable driver's seat. Not available with articulating bucket seats or leather trim groups.	
	Firebird coupe with 1SA:	305.00
	Firebird coupe with 1SB or 1SC:	270.00
	Convertibles:	270.00
UK3	Steering wheel radio controls. With leather appointment group. Includes leather-wrapped steering wheel and gear shift lever. Standard Trans Am convertible. Included in 1SC on coupes and 1SB on convertibles. Not available with 1SA on Firebird or Formula coupes. Requires radio upgrade if ordered on Firebird or Formula coupe.	
	Firebird and Formula:	200.00
	Trans Am coupe:	125.00
NW9	Traction control system. Not available on base Firebird. Requires QCB or QFZ tires.	450.00
MXO	Four-speed automatic transmission. Not available with 1SA on Firebird coupe.	790.00
T43	Up-level rear deck lid spoiler. Not available on Firebird or Formula models. Trans Am coupe:	395.00
PQ5	Chromed aluminum wheels. Not available with 1SA on Firebird coupe.	500.00

1997

★	Firebird Convertible
★	Formula Coupe
★★	Formula Convertible
★★	Trans Am Coupe
★★★	Trans Am Convertible
★★★★	Firebird Firehawk Coupe
★★★★★	Firebird Firehawk Convertible

Note: Add one star to ratings for Formula and Trans Am Coupes or Convertibles with the WS6 Pontiac Ram Air Performance Package.

"Either you get it or you don't," was one of the slogans used to sell '97 Firebirds. The simple message was blunt and clear. Even the base Firebird was a real eye-catching car that got appreciative looks as it drove down the street. You could get a Firebird in a variety of flavors, from afford-able to expensive, from economical to guzzler, and from sports car to muscle car, but you couldn't get a Firebird that would please everyone. It was still a car for young, successful driving enthusiasts who wanted bold, aggressive styling.

The typical buyer of a V-6 Firebird coupe was a 36-year-old person making $55,000 a year. Just over half were married, and just below half were male college graduates working in a professional field. At the other end of the Firebird spectrum, Trans Am buyer demographics indicated a median age of 40, average household income of $75,000, and that slightly over 50 percent of buyers were married males with a college degree and professional occupation.

Air conditioning and the PASS-Key II theft-deterrent system became standard equipment for all 1997 Firebirds. Also standard was a four-way seat adjuster located on the left-hand side of the driver's seat.

Starting this year, all Firebird convertibles included a number of standard features, including power mirrors; power windows; power door locks; cruise control; and rear-seat courtesy lamps. The ragtops included a flush-folding power top that stored beneath an easily-assembled three-piece hard tonneau cover. When the top was raised, the standard rear-glass window with electric defogger helped give a clear view out

The Firebird looked much the same, but got a number of equipment upgrades for 1997. One was the optional Sport Appearance Package seen on this coupe. It made the low-end car look Trans Am-like.

the back of the car. And a fully-trimmed headliner contributed to a quieter ride.

Body-side moldings were standard on all models now, and at midyear, all Firebirds got windshield wipers that went better with Firebird's sleek design. Bright green metallic (code 31) was one new exterior paint choice and bright purple metallic (code 88) was added at midyear. Other colors were bright white (code 10); bright silver metallic (code 13); black (code 41); dark green metallic (code 48); blue-green chameleon (code 79); bright blue metallic (code 80); bright red (code 81); and red-orange metallic (code 96). Daytime running lamps (DRLs) were made standard. Unlike conventional DRLs, which use low-intensity regular headlamps, the Firebirds used special high-intensity elements in the parking-signal lamps.

New Firebird interior features included a redesigned center console with an auxiliary power outlet for cell phones and other electronic devices. The console came with a pull-out side cup holder and a revised storage area. Dark Pewter interior finish was newly added, along with Cartagena cloth upholstery. Heating, ventilation, and air-conditioning (HVAC) control knobs were back-lighted for better nighttime visibility. Also new inside was a 500-watt peak power Monsoon sound system.

The 3.8-liter V-6 had a new vibration dampener to reduce vibrations in the engine compartment at high rpms. The base engine produced 200 horsepower. Available again was the 3800 V-6 Performance Package, which included four-wheel disc brakes; a limited-slip differential; dual-outlet exhausts; five-spoke cast aluminum wheels (also available in chrome); P235/55R16 tires; a 14.4:1 steering ratio; and 3.42:1 axle with

Starting this year, all Firebird convertibles—even this base model—came with power mirrors; power windows; power door locks; cruise control; and rear-seat courtesy lamps.

The Formula was basically a Firebird V-8 with a fat rear stabilizer bar, quick steering, and four-wheel disc brakes. This coupe has the base 285-horsepower, 5.7-liter V-8 under the standard-design hood.

automatic transmission. There was also a W68 Sport Appearance Package for V-6s, which added or substituted body ground-effects components; specific fog lamps; and dual-outlet exhausts with cast-aluminum extensions.

The Formula was basically a V-8-powered Firebird with a fatter rear stabilizer bar, a quicker steering ratio, and four-wheel disc brakes. The standard 5.7-liter V-8 produced 285 horsepower at 5,000 rpm and 325 foot-pounds of torque at 2,400 rpm. With the WS6 Ram Air Package, these ratings changed to 305 horsepower at 5,400 rpm and 335 foot-pounds at 3,200 rpm. Cars with the $3,000 WS6 package were again characterized by the use of a twin-nostril hood with a Ram Air logo on each air-inlet opening.

The Trans Am was basically a customized Formula. Its distinct fog-lamp front fascia added 1.4 inches to its overall length, and 1.2 inches to its front overhang. It also had a fraction-of-an-inch more overhang at the rear. Trans Am coupes also included power Sport mirrors, cruise control, power door locks, an electric rear-window defogger, leather-wrapped interior, and power windows. Trans Am convertibles also had a power antenna; P245/50ZR16 all-weather speed-rated tires; a Delco 2001 series stereo; remote-keyless entry; and a leather-wrapped interior with steering-wheel radio controls. Formulas said nothing on the sides of the car, but Trans Ams had the model name on the side, as did base Firebirds.

Pontiac expanded availability of the WS6 Performance & Handling package to V-8-powered Formula and Trans Am convertibles. This

In 1997, Pontiac expanded availability of the WS6 Performance & Handling package to V-8-powered Formula and Trans Am convertibles. The Ram Air Trans Am ragtop is seen here.

The Trans Am was a customized Formula. The front and rear fascias made it slightly longer. For 1997, coupes had power mirrors, locks, and windows; cruise control; an electric rear-window defogger; and leather-wrapped goodies.

Ragtops included a flush-folding power top; an easily-assembled three-piece hard tonneau cover; a rear-glass window with electric defogger; and a fully-trimmed headliner for a quieter ride.

instantly created the year's two most collectible models, as these rapid ragtops seemed to have it all, from looks, to go-power, to rarity.

Pontiac Motor Division built an intriguing prototype car in its engineering garage during 1997. It was a hatch-roof coupe with silver paint, a special American Sunroof Corp. (ASC) graphics package, and a Ram Air version of the 3800 Series II V-6. *High-Performance Pontiac* magazine reported positively on the car's performance in its August 1997 issue, but said that PMD was not saying whether it had plans to do a production version.

Meanwhile, our friends at SLP (Street Legal Performance), in Tom's River, New Jersey, planned to build about 8,000 cars with Firehawk-type performance conversions. However, this approved aftermarket "tuner" had started doing both Firebird and Camaro conversions in 1996, which means that not all of the 1997 SLP models are Firebirds. Nevertheless, GM continued to farm out specialty work to companies like ASC and SLP, and this results in some very interesting vehicles with instant collector-car status.

Left: New-for-'97 interior features included a redesigned center console, dark pewter interior finish, Cartagena cloth upholstery, back-lighted HVAC controls, and a new 500-watt peak power Monsoon sound system.

1997 Firebird Models and Base Prices

Model Number	Model Name	Dealer Cost	Suggested Retail Price
S87S	Firebird Coupe	$15,714	$17,174
S67S	Firebird Convertible	$21,122	$23,084
V87S	Formula Coupe	$18,962	$20,724
V67S	Formula Convertible	$24,269	$26,524
V87S, Y82	Trans Am Coupe	$20,939	$22,884
V67S, Y82	Trans Am Convertible	$26,026	$28,444

Dealer destination charge (all models effective Jan. 2, 1997):$525

Basic standard equipment: 3.8-liter fuel-injected V-6; power front disc brakes; five-speed manual transmission; and P215/60R16 black sidewall touring tires.

1997 Firebird Options and Accessories

UPC Code	Description	Retail Price ($)
Emission Systems		
FE9	Federal	n/c
YF5	California (Base Firebird)	n/c
	Formula, Trans Am	170.00
NG1	New York and Massachusetts (Base Firebird)	n/c
	Formula, Trans Am	170.00
NB8	Override, Calif./N.Y./Mass. Requires FE9	n/c
n/c7	Override, Federal. Requires YF5 or NG1	n/c
Wheels & Tires		
PO5	Chromed 16-inch aluminum wheels	595.00
QCB	P235/55R16 Blackwall. Touring. STL. Standard: Formula. Not available on Trans Am.	132.00
QFZ	P245/50ZR16 Blackwall. STL. All-weather performance. Standard: Trans Am convertible. Formula coupe and convertible, Trans Am coupe:	245.00
QLC	P245/50ZR16 Blackwall. STL. High performance. Not available on base Firebird or with NW9 Traction Control. Included with WS6 package on Formulas and Trans Ams:	245.00
QFK	P275/40ZR17 speed-rated tires. Available with WS6 only Formula, Trans Am coupes:	n/c
Interiors		
AN3	Articulating bucket seats. Includes adjustable lumbar support. With Prado leather trim. Firebird, Formula:	804.00
AQ9	Includes AN3, plus articulating headrests	
	Coupe (Requires W54 or W56 radio upgrade):	829.00
	Convertible:	804.00
W68	Sport Appearance Package. Includes specific aero appearance; fog lamps; dual outlet exhausts. On base Firebirds requires QCB tires. Not available with NM5.	1,449.00
Y87	3800 performance package. Includes limited-slip differential; 4-wheel disk brakes; up-level steering; dual outlet exhaust; and QCB tires. Also includes 3.42:1 rear axle ratio if ordered with MXO transmission. Available on Firebird coupe and convertible only:	550.00
1LE	Performance Package. Includes special-handling suspension with larger stabilizer bars; stiffer springs; and Koni shock absorbers. Requires WS6 Ram Air package and MN6 manual transmission. Not available with MXO transmission. Free standing options limited to C49 defogger and VK3 license plate bracket. Formula coupe with 1SA:	1,175.00

UPC Code	Description	Retail Price ($)
WS6	Ram Air performance and handling package. Includes Ram Air induction system; specific tuned suspension; five-spoke 17-inch aluminum wheels; P275/40ZR17 speed-rated tires; and dual oval high-polished exhaust outlets. Formula, Trans Am coupes:	3,345.00
WS6	Ram Air performance and handling package. Includes Ram Air induction system; five-spoke 16-inch aluminum wheels; P245/50ZR16 high-performance tires; and dual oval high-polished exhaust outlets. Formula, Trans Am convertibles:	2,995.00
UA6	Content theft alarm. Requires AUO keyless entry. Not available with 1SA on Firebird coupe.	90.00
K34	Cruise control with resume speed on coupes. Standard: Trans Am.	235.00
C49	Rear-window defogger on coupes. Standard: Trans Am.	180.00
CC1	Removable hatch roof with locks and stowage. Coupes only. Not available with 1SA on Firebird nor Formula.	995.00
VK3	Front license plate bracket	n/c
DG7	Left- and right-hand power mirrors with blue glass, on coupes. Standard: Trans Am.	96.00
U75	Power antenna. Requires optional radio upgrade if ordered on Firebird or Formula coupe. Standard: Trans Am convertible.	85.00
AU3	Power door locks, on coupes. Standard: Trans Am coupe, all convertibles.	220.00
AG1	6-way power adjustable driver's seat for coupes. Not available with articulating bucket seats nor leather trim groups.	270.00
A31	Power windows, on coupes. Standard: Trans Am.	290.00

Sound Systems

UPC Code	Description	Retail Price ($)
W53	ETR AM/FM stereo with CD player; graphic equalizer; clock; seek up/down; and four-speaker coaxial sound system. Available in coupes only.	100.00
W54	"Monsoon" ETR AM/FM stereo with auto reverse cassette; graphic equalizer; clock; seek up/down; remote CD prewiring; and 10-speaker sound system.	
	Coupes with 1SA:	230.00
	Coupes with 1SB:	130.00
W55	ETR AM/FM stereo with CD player; graphic equalizer; clock; seek up/down; 10-speaker sound system.	
	Coupe with 1SA:	330.00
	Coupe with 1SB:	230.00
W73	ETR AM/FM stereo with CD player; graphic equalizer; clock; seek up/down; 6-speaker sound system. Available in convertibles only.	100.00
U1S	Remote 12-disc CD changer, trunk mounted in coupes. Requires W52 or W54 radio upgrade.	595.00
AUO	Remote keyless entry. Standard: Trans Am convertible.	150.00
GU5	Performance rear axle. Not available on Firebird.	225.00
UK3	Steering wheel radio controls. With leather appointment group. Standard: Trans Am convertible.	
	Firebird and Formula:	200.00
	Trans Am coupe:	125.00
NW9	Traction Control system. Not available on base Firebird. Requires QCB or QFZ tires.	450.00

Transmissions

UPC Code	Description	Retail Price ($)
MN6	Six-speed manual. Not available on base Firebird. Formula, Trans Am:	n/c
MM5	Five-speed manual. Standard and only available on base Firebird coupe. No charge with 1SA. On coupe with 1SB, the following amount is a credit for deleting the four-speed automatic transmission.	(815.00)
MXO	Four-speed automatic. Standard: Formula and Trans Am. Firebird:	815.00

1998

For 1998, Pontiac's stable of aggressively-styled cars includes a bold new Firebird. In addition to a new appearance guaranteed to live up its legendary status among sports-car enthusiasts, the 1998 Firebird has numerous functional improvements, including more V-8 horsepower and torque for Formula and Trans Am models.

The 1998 base Firebird and Formula models share one new front fascia design, while the Trans Ams have a different new design. Both styles feature center twin ports below the hood, and both have restyled round fog lamps. On Trans Ams, the fog lamps are located near the center, while the Firebird and Formula fog lamps are located further outboard. Two new paint colors, navy metallic and sport gold metallic are available.

All major elements of the 1998 front end, including the hood, fascia, fenders, and headlamps, are new. The aggressive-looking hood design provides smoother contours, and also improves underhood airflow. The standard hood

★	**Firebird Convertible**
★	**Formula Coupe**
★★	**Trans AM Coupe**
★★★	**Trans Am Convertible**
★★★★	**Firebird Firehawk Coupe**
★★★★★	**Firebird Firehawk Convertible**

Note: Add one star to ratings for Formulas and Trans Ams with the WS6 Pontiac Ram Air Performance Package.

has two openings with a "Ram Air" look that are actually non-functional. A revised 320-horsepower Ram Air (WS6) package did arrive at midyear. The Ram Air nose has two large "ant eater" air scoops that are divided, in front, by a horizontal fin or blade. This is a very, very racy-looking design. Small "Ram Air" lettering appears on the sides of the scoops.

The new pop-up headlamps are of a "mini-quad" design that enhances appearance of the car when they are in use, and also gives significantly improved road lighting. GM wanted Pontiac to drop flip-up headlights, because they're costly, but market researchers determined that they were a trademark of the Firebird that's burned into the public's mind, so they stayed. At close distances, the new lighting pattern is wider, and at longer distances, the intensity of the lights is greater than before.

The Firebird's aggressive new fender design, with extractors (or brake-cooling ducts), is common to all models. A new 16-inch wheel is used on cars with the base V-8. Optional chrome wheels are carried over from 1997, as are the 17-inch wheels on Ram Air cars.

Formula and Trans Am models also have new rear end styling. The sales brochure says that their "new taillamps will be a familiar sight." Inside were gauges with clear white characters on black, analog faces, to help keep drivers informed of what was going on with their cars.

Mechanical upgrades include larger brakes all around. Up front, dual-piston aluminum calipers replace the old cast iron, single-piston jobs. The parking brake is reworked, too. Organic pads replace the previous semi-metallic pieces, and the

The base Firebird had a new front fascia, which was shared with Formulas. The round fog lamps were standard, and further outboard on these models. The front fender air extractors harken back to early Trans Ams.

This 1998 Formula coupe sports the new WS6 "Ant-eater" Ram Air hood and 320-horsepower performance package. This was a midyear addition to the Firebird options list. There was no more Formula ragtop.

ABS hardware is also improved with a new Bosch solenoid.

Standard equipment for the 1998 base Firebirds includes extensive acoustical insulation; dual air bags; air conditioning; a black fixed-mast antenna (mounted at the right rear); a brake-transmission shift interlock (with automatic transmission only); four-wheel power-disk ABS brakes; cruise control; electric rear- and side-window defoggers; 3.8-liter V-6; Solar-Ray tinted glass; electronic-analog instrumentation; LED trip odometer; dual Sport mirrors (left-hand remote-controlled); day/night rear view mirror with reading lamps; covered left- and right-hand visor-vanity mirrors; a Delco 2001 ETR AM/FM stereo with CD player (including a seven-band equalizer, clock, touch control, seek up/down, search-and-replay, Delco theft lock, remote CD prewiring and four-speaker coaxial sound system); reclining front bucket seats; folding rear seat; four-way manual driver's seat adjustment; four-spoke tilt steering wheel; adjustable steering column; theft-deterrent system; PASS-Key II system; P215/60R15 steel-belted radial black-sidewall touring tires; five-speed manual transmission; controlled-cycle windshield wipers; and 16-inch bright silver, five-spoke cast-aluminum wheels. (Whew!)

As you can see on this coupe, the Trans Am parking lamps, though also round-shaped, are positioned more towards the center of the front fascia. Retractable headlamps were retained, despite higher production cost.

Firebird Formulas add or substitute a power antenna; power door locks; dual power Sport mirrors with blue glass; a 500-watt peak-power "Monsoon" radio with CD and seven-band graphic equalizer (includes clock, touch control, seek up/down, search-and-replay, Delco theft lock, and high-performance 10-speaker system for coupes or six-speaker system in convertibles). The sound system includes 6.5-in HSS speakers and tweeters in the doors, 6.5-inch subwoofers in the sail panels, a sub-woofer amp, and four-inch speakers and tweeters in the rear quarter panels (except on convertibles.) The system requires power windows and door locks, the leather appointments group, P235/50ZR16 speed-rated all-weather tires; and power windows with an express-down feature (requires power windows and door locks).

Trans Ams came with all this plus an audible theft-deterrent system, Remote Keyless entry, and a six-way power front driver's seat.

The beefed-up LS1 5.7-liter V-8, which replaced the LT1, generated 305 horsepower at 5,200 rpm and 335 foot-pounds of torque at 4,000 rpm for enhanced mid-range responsiveness. That's 20 more horsepower and 10 more foot-pounds of torque than the previous Firebird V-8. The WS6-equipped Ram Air LS1 V-8 boasted 320 horsepower. A "modern muscle car" Pontiac calls it. The base Firebirds keep using the high-torque, 200-horsepower 3800 Series II V-6 engine, which provides low operational costs. The WG8 Sport Appearance Package and 3800 V-6 Performance Package were available, too.

In October 1997, *High-Performance Pontiac* magazine featured the 1988 Firebird on its cover and in a story titled "Perfecting Excellence." It explains the LS1 engine improvements in great detail and gives positive seat-of-the-pants impressions of the performance of the '98 over the already-hot '97. The latest Firebird may be the greatest one. It is at least definitely in the same league as the muscle cars of yesteryear.

1998 Firebird Models and Base Prices

Model Number	Model Name	Dealer Cost	Suggested Retail Price
S87S	Firebird Coupe	$16,484	$18,015
S67S	Firebird Convertible	$22,239	$24,305
V87S	Formula Coupe	$20,921	$22,865
V87S, Y82	Trans Am Coupe	$23,767	$25,975
V67S, Y82	Trans Am Convertible	$27,189	$29,715

Dealer destination charge (all models effective Sept. 22, 1997): $525

1998 Firebird Options and Accessories

UPC Code	Description	Retail Price ($)
Option Groups		
Firebird Coupe		
1SA	Includes vehicle standard equipment only	n/c
1SB	Includes 1SA plus power door locks; power windows; dual power mirrors; power antenna; power seat	1,610.00
1SC	Includes 1SB plus power seat; security package with theft deterrent system; and remote keyless entry.	2,450.00
	Formula Coupe	
1SB	Includes power seat; security package with theft deterrent system; remote keyless entry system; and removable hatch roof	1,505.00
	Firebird Convertible	
1SA	Includes security package with theft deterrent system; and remote keyless entry system	n/c
	Trans Am	
1SA	Includes same as Firebird convertible	n/c
VK3	Front license plate bracket	n/c
Emission Systems		
FE9	Federal	n/c
YF5	California	170.00
NG1	New York, Massachusetts, Connecticut	170.00

UPC Code	Description	Retail Price ($)

Other Options

GU5	Performance rear axle. Not available on Firebird.	225.00
NW9	Traction control system. Not available on base Firebird. Requires QCB or QFZ tires.	450.00
WS6	Ram Air performance and handling package	3,100.00
Y87	3800 performance package	440.00
PO5	Chromed 16-inch aluminum wheels	595.00

Sound Systems

W54	"Monsoon" ETR AM/FM stereo with auto reverse cassette; graphic equalizer; clock; seek up/down; remote CD prewiring; and 10-speaker sound system.	
	Coupes with 1SA:	230.00
	Trans Am coupes:	(100.00)

The Trans Am convertible is about as cool as you can get. Unless you bought a Firehawk convertible from SLP Engineering, that is.

		Retail Price ($)
W55	ETR AM/FM stereo with CD player; graphic equalizer; clock; seek up/down; 10-speaker sound system.	
	Firebird coupes with 1SC:	n/c
	Other Firebird coupes:	430.00
W59	ETR AM/FM stereo with auto-reverse cassette; graphic equalizer; clock; seek up/down; remote CD prewiring and six speakers. Convertible:	(100.00)
CC1	Removable hatch roof with locks and stowage. Coupes only. Included on Formula coupe with 1SB.	995.00
R7X	Security package for coupe. Included with Firebird 1SC or Formula 1SB.	240.00

Interiors

AQ9	Articulating custom bucket seats with adjustable lumbar supports	155.00
AR9	Leather trimmed front bucket seats	650.00
AG1	Six-way power driver's seat in coupes. No charge with Firebird with 1SB or Formula with 1SC.	270.00
W68	Sport Appearance Package	990.00

Transmissions

MN6	Six-speed manual. Not available on base Firebird. Formula, Trans Am:	n/c
MM5	Five-speed manual. Standard and only available on base Firebird coupe. No charge with 1SA. On coupe with 1SB/1SC, the following amount is a credit for deleting the automatic ransmission.	(815.00)
MXO	Four-speed automatic. Standard: Formula, Trans Am. No charge in base Firebird with 1SB, 1SC. In other base Firebirds:	815.00

Sources

ALABAMA
Donnie Hodges
Rt. 3, Box 366
Albertville, AL 35950
(parts)

Midland Automotive Products
33 Woolfolk Ave.
Midland City, AL 36350
(molded carpets)

ARIZONA
Phoenix Firebird Parts
4101 E. Karen Dr.
Phoenix, AZ 85032
(parts, emblems)

CALIFORNIA
B & M Automotive Products
9152 Independence Ave.
Chatsworth, CA 91311

Herb Adams VSE
125 Ocean View Blvd.
Pacific Grove, CA 93950
(specialty Trans Am models)

Hooker Headers
1032 W. Brooks St.
Ontario, CA 91761
(Trans Am bolt-on accessories)

H.O. Racing Specialties
P.O. Box 429 PG
Hawthorne, CA 90250
(speed equipment)

HT Enterprises
P.O. Box 4362
Santa Rosa, CA 95402
(structural fender connectors for Trans Am/Camaro)

Lamm-Morada Publishing Co.
Box 7607
Stockton, CA 95207
(books)

Pacific T-Top, Inc.
15241 Tansistor Lane
Huntington Beach, CA 92649
(accessories)

Paddock West
446 Tennessee St.
Redlands, CA 92373
(sheet metal, interiors, accessories)

Peter McCarthy Enterprises
P.O. Box 664
Temple City, CA 91780
(Super-Duty parts catalog)

FLORIDA
Robinson's Collector's Corner
1417 South St.
Leesburg, FL 32748
(sales literature, tech manuals)

The Pontiac Store
P.O. Box 162103
Miami, FL 33116
(T-shirts)

Crestline Publishing
1251 N. Jefferson Ave.
Sarasota, FL 33577
(books)

GEORGIA
Jim Osborn Reproductions
3070 D Briarcliff Rd. NE
Atlanta, GA 30329
(decal reproductions)

Year One
Box 450131
Atlanta, GA 30345
(1969 Trans Am fiberglass reproduction
parts, stock and custom body parts)

ILLINOIS
Mr. Chrome
319 N. Whipple St.
Chicago, IL 60612
(new and used bumpers)

Ron Fulsang
8246 S. 85th
Justice, IL 60458
(postcards)

INDIANA
Dick Choler
640 E. Jackson St.
Elkhart, IN 46514
(parts, cars)

The Paddock Inc.
38 W. Warrick St., Dept. PG
P.O. Box 30
Knightstown, IN 46148
(panels, parts, accessories, reproductions)

Triple A Enterprises
P.O. Box 50522
Indianapolis, IN 46250
(dealer invoice reproductions)

KENTUCKY
Bob Cook Classic Parts
Dept. CX
Hazel, KY 42049
(windshields)

MARYLAND
Bob's Pontiac Parts
Box 333
Simpsonville, MD 21150
(interior parts)

MASSACHUSETTS
Classic Glass
287 Salem St.
Woburn, MA 01801
(curved glass)

Dave Edwards
Box 245-OC
Needham Hts, MA 02194
(Hydra-matic transmission parts)

MICHIGAN
Harry Samuel
65 Wisner St.
Pontiac, MI 48058
(molded carpets)

Roger Lee
17134 Wood St.
Melvindale, MI 48122
(parts)

J & J Sheldon
2718 CE Koper St.
Sterling Heights, MI 48077
(radios, sound equipment through 1972)

Mid-America
238 Wayne Rd.
Westland, MI 48185
(high-performance engine hardware)

Sherman & Associates
P.O. Box 644
St. Clair Shores, MI 48080
(steel patch panels)

C.A.R.S., Inc.
1102 Combermere St.
Troy, MI 48084
(1969 inner fender panels)

MINNESOTA
Jere V Longrie
Rt. 4, Box 8884
Grand Rapids, MN 55744
(sales literature, tech manuals)

MISSOURI
Bob's Pontiac Parts
Box 333
Simpsonville, MO 21150
(sheet metal, literature, emblems,
upholstery, rubber parts, decals)

NEBRASKA

Auto Krafters
Box 9
Fort Calhoun, NE 68023
(books)

Edward Rohrdanz
3333 Melrose Ave.
Lincoln, NE 68506
(literature, ads)

NEW HAMPSHIRE

Ames Performance Engineering
Bonney Rd.
Marlborough, NH 03455
(parts, emblems, literature, decals,
patch panels)

NEW JERSEY

Fatsco Transmission
81 Route 46
Fairfield, NJ 07006
(Hydra-matic parts and service)

A.J. Speed Equipment
1290 Liberty Ave.
Hillside, NJ 07205
(custom repairs)

Trans Am Specialties
1514 Route 38
Cherry Hill, NJ 08034
(performance car dealer/builder
manufacturer of Bandit Trans Am)

Vintage Parts House
93 Whippany Rd.
Morristown, NJ 07960
(carpets)

Media Marketing Services
13 Brentwood Rd.
Matawan, NJ 07747
(books)

Quicksilver Communications
Performance Book Div.
167 Terrace St.
Haworth, NJ 07641
(books)

NEW YORK

High-Performance Pontiac
McMullen Argus Publishing, Inc.
774 S. Placentia Ave.
Placentia, CA 92870-6846
(specialized Pontiac magazine)

Matt Mariacher
Elma, NY 14059
716-674-5496
(parts)

Motion Performance Parts
598 Sunrise Hwy.
Baldwin, NY 11510
(fiberglass hoods, spoilers, flares)

Nunzi's Automotive
Nunzi Romano
1419 64th St.
Brooklyn, NY 11219
(Pontiac engine modifications)

Trans Am Headquarters
1840 E. Tremont Ave.
Bronx, NY 10460
(specialty dealer & body shop)

Walter Miller
315 Wedgewood Terr.
DeWitt, NY 13214
(sales literature, tech manuals)

NORTH CAROLINA

Marty McDaniel
P.O. Box 1175
Hickory, NC 28601
(parts, parts listing newsletter)

OHIO

Dick Kreiger
2451 Patterson Rd.
Dayton, OH 45420
(sheet metal, literature, emblems)

Mainstream
P.O. Box 07403
Columbus, OH 43297
("chicken" hood decal sets)

PENNSYLVANIA

Reden's Relics
P.O. Box 300
Rolling Springs, PA 17007
(parts, literature)

Auto Accessories of America
Box 427, Rt. 322
Boalsburg, PA 16827
(parts)

J. Coeyman Enterprises
P.O. Box 61
Dallastown, PA 17313
(performance axles)

OREGON

Associated Car Clubs
235 E. Third St.
McMinnville, OR 97128
(books, literature, ads)

TEXAS

Ron's Pontiac Parts
Box 1527
Sherman, TX 75090
(stock and performance V-8 parts)

WASHINGTON

Mecham Racing Inc.
8200 Sprague Ave. South
Tacoma, WA 98409
(Trans Am SCCA race car builder)

WISCONSIN

MBI Publishing Company
P.O. Box 1
Osceola, WI 54020
(literature)

Doug LaKosh
P.O. Box 103
Genesee Depot, WI 53127
(Pontiac parts)

Larry Lorenz
6107 N. 118th St.
Milwaukee, WI 53225
(parts, literature)

Parts of the Past
P.O. Box 602
Waukesha, WI 53187
(Trans Am literature)

Old Cars Newspaper
700 E. State St.
Iola, WI 54990
(weekly tabloid)

Bibliography

BOOKS

Bonsall, Thomas E. *Firebird: A Source Book.* Bookman Dan, Baltimore, MD.

Bonsall, Thomas E. *Pontiac: The Complete History.* Bookman Dan, Baltimore, MD.

Clarke, R. M. *Pontiac Firebird.* Brooklands Books, London, England.

Gunnell, John A. *75 Years of Pontiac-Oakland.* Crestline Publishing, Sarasota, FL.

Norbye, Jan and Jim Dunne. *Pontiac: The Postwar Years.* Motorbooks International, Osceola, WI.

Oldham, Joe. *Supertuning Your Firebird Trans Am.* Tab Books, Blue Ridge Summit, PA.

Schorr, Martyn L. *Pontiac: The Performance Years.* Quicksilver Publications, Hawthorne, NJ.

Schorr, Martyn L. *Pontiac Trans Am: America's Premier Pony Car.* Quicksilver Publication, Hawthorne, NJ.

Witzenberg, Gary. *Firebird Trans Am: Amercia's Premier Performance Car.* Automobile Quarterly Publications, Kutztown, PA.

Index